LOSING A LOST TRIBE

LOSING A LOST TRIBE

Native Americans, DNA, and the Mormon Church

Simon G. Southerton

Signature Books | Salt Lake City | 2004

SIMON G. SOUTHERTON is a senior research scientist with the Commonwealth Scientific and Industrial Research Organization (CSIRO) in Canberra, Australia. He is a former senior research scientist in the Department of Biochemistry, University of Queensland, and postdoctoral fellow at the John Innes Centre in Norwich, England. He holds a Ph.D. from the University of Sydney in plant science and now specializes in the molecular biology of forest trees. He has published research articles in international journals such as *Plant Molecular Biology, Plant Physiology*, and *Physiological and Molecular Plant Pathology*. He served an LDS mission to Melbourne in the 1980s.

Cover design: Julie Easton

Losing a Lost Tribe was printed on acid-free paper and was composed, printed, and bound in the United States of America.

See: http://www.signaturebooks.com

09 08 07 06 05 04 6 5 4 3 2

LIBRARY OF CONGRESS CATALOGING-IN-PUBLICATION DATA
Southerton, Simon G.
Losing a lost tribe : Native Americans, DNA, and the Mormon Church / by Simon G. Southerton.
p. cm.
Includes bibliographical references (p.) and index.
ISBN: 1-56085-181-3 (pbk.)
1. Book of Mormon. 2. Indians–Origin. 3. Lost tribes of Israel. I. Title.

BX8627.S647 2004
289.3'2–dc22

2004052191

Contents

Appendices

Preface

The Mormon belief that native people in the Americas and Polynesia are largely descended from Israelites is fundamental to the Church of Jesus Christ of Latter-day Saints, the "LDS" or Mormon church. In addition, Mormons believe that these wandering Hebrews, when they arrived in the New World, were practicing Christians centuries before the birth of Christ. These views come from the sacred writings Mormons possess in addition to the Bible, in particular the Book of Mormon. Members of the faith firmly proclaim special knowledge in this area of anthropology and have long resisted other evidence to the contrary. Oblivious to territorial boundaries, scientists also form views about human origins and the routes our various ancestors took when they colonized the earth. Consequently, the stage has long been set for a debate, and both sides have accumulated evidence they find compelling. I will try as best as I can to introduce readers to both sides and to offer my views on what I consider to be the inevitable conclusion to the matter.

Few topics polarize opinion as sharply as the interface of religion and science, which is why frank or objective discussion in these areas is difficult. However, through the medium of books, we can speak to each other with enough distance to allow some perspective, and we can contemplate one another's thoughts in the safety of our own homes. In this book, I will use the moderating tone of third-person references to both scientists and Mormons. Disinterested commentators are rarely sufficiently motivated to write, and I am no exception. Several years ago I encountered research into molecular genealogy that compelled me to compare what I thought I knew religiously with what I knew from my training in science. I now have strong opinions about the topic, which I will express freely, as will be quite obvious to readers. My wish is not to offend or to offer advice in matters of faith. These are issues that each person has to decide on his or her own. But for fellow Mormons who believe American Indians and Polynesians are largely descended from ancient Israelites, the recent findings of science may compel them, as I was compelled, to re-evaluate their thinking.

For the sake of disclosure, I was a fully active member of the Mormon church for almost thirty years. I served a two-year Mormon mission, married in an LDS temple, and served in several teaching and leadership positions including two years as a bishop. Like all Mormon bishops, I held down a full-time day job. Professionally, I have spent most of my career researching in the field of molecular plant biology. The plant and animal kingdoms are closely related on a molecular level, and the structure and language of DNA are remarkably conserved in both. Just as molecular biology helps researchers decode human origins, it is applied to other animals and to plants to unravel genetic relationships and evolutionary origins.

A note about terms. I will sometimes refer to Latter-day Saints as simply "the Saints," as members of the church often refer to themselves as a convenient shorthand reference. I have already mentioned the other forms of reference—LDS and Mormon. In addition, I will employ the term Gentile, which is how LDS people refer to non-Mormons. Latter-day Saints regard themselves as the adopted House of Israel and see anyone who is neither Mormon nor Jewish as a Gentile.

The church hierarchy includes a president who is revered as a prophet of God. He and two counselors act together as the First Presidency. Subordinate to the presidency is the Quorum (or Council) of the Twelve Apostles, twelve men who are considered to be "prophets, seers and revelators," ranked in the order of the date of their ordination to the apostleship. The senior apostle traditionally becomes the prophet upon the death of the previous incumbent. Occasionally these men are referred to as "the Brethren." The next level of the hierarchy includes members of the Quorum (or Council) of Seventy, many of whom occupy positions in area presidencies throughout the world. All of the above positions are filled by men, and they are collectively known as the "General Authorities."

I will refer to LDS scriptures throughout, so a list of the books within the Book of Mormon may be helpful, shown here in sequential order with the officially recognized abbreviations in parentheses:

First Book of Nephi (1 Ne.)
Second Book of Nephi (2 Ne.)
Book of Jacob (Jacob)
Book of Enos (Enos)
Book of Jarom (Jarom)
Book of Omni (Omni)
The Words of Mormon (WofM)
Book of Mosiah (Mosiah)
Book of Alma (Alma)
Book of Helaman (Hel.)
Third Nephi (3 Ne.)
Fourth Nephi (4 Ne.)
Book of Mormon (Morm.)
Book of Ether (Ether)
Book of Moroni (Moro.)

Other uniquely LDS scriptures include the Doctrine and Covenants (D&C) and the Pearl of Great Price. The latter includes the following:

Book of Moses (Moses)
Book of Abraham (Abr.)
Joseph Smith—Matthew (JS-M)
Joseph Smith—History (JS-H)

The church otherwise accepts the authorized King James Version of the Bible.

I am grateful for the assistance of several scientists involved in New World and Pacific studies including Theodore Schurr, Andrew Merriwether, Antonio Torroni, David Glenn Smith, and Peter Bellwood. Most of these gentlemen are, by-and-large, unaware of the theological implications of their research, but they happily shared research findings and personal observations and opinions. I would also like to thank the following for critically reading various stages of the manuscript: Gary Edmond, Thomas Murphy, Richard Packham, Bob Birks, Nigel Wace, Kevin Thompson, Romi Thompson, Steven Clark, Jane Southerton, and some LDS friends who prefer to remain anonymous. I would like to express thanks to my wife and family who have patiently awaited completion of this book. While I have benefitted from the extensive work and assistance of many people, I take full responsibility for this book and whatever errors persist despite my best efforts to eliminate them.

Introduction

And thy seed shall be as the dust of the earth, and thou shalt spread abroad to the west, and to the east, and to the north, and to the south: and in thee and in thy seed shall all the families of the earth be blessed.

—Genesis 28:14

When Christopher Columbus launched into the Sea of Darkness over 500 years ago, his intention was to find a quick route from Spain to the riches of the Indies. Relying upon Ptolemy's (AD 90-168) maps of a spherical earth bearing only the continents of Europe, Asia, and Africa, Columbus was certain he had achieved this goal when he landed in the Bahamas in 1492. It was about where he had expected the Indies to be. All his life he stubbornly refused to accept that he had discovered a new continent—the world he knew having had no room for a western hemisphere. His error was due to calculations of the earth's circumference that made the earth a quarter too small. Centuries later, Europeans came to grips with the geography of the quarter of the earth that had eluded Ptolemy's pen. This miscalculation lives on in the popular misunderstandings about the origins and diversity of native people who inhabit that geography.

Europeans were at first mystified by the presence of people at such great distances from the centers of civilization that were familiar to the Judeo-Christian world. Not surprisingly, early attempts to account for their origins were ensnared in the biblical mindset, the widely accepted worldview among members of the European societies that emerged from the ashes of the Roman Empire. In many cases the native inhabitants of the Americas and the Pacific were regarded as savages, the degraded remnants of once civilized nations whose origins could be traced back to Noah's offspring. A common and persistent theory among early Europeans was that Native Americans and Pacific Islanders were the scattered remnants of the House of Israel.

For over a century, the vast majority of scholars and scientists have been satisfied that Native Americans and Pacific Islanders share a common ancestry. But it is not in Israel. The academic world has accumulated a comprehensive library of work that links each of these groups with an ancient homeland in Asia. Most scholars now accept that the ancestors of the American Indians began migrating to the Americas from somewhere in the vicinity of southern Siberia, across an icy Bering Strait, over 14,000 years ago. Similarly convincing are the signs that the early colonizers of the Pacific Isles began emerging from Southeast Asia about 30,000 years ago. The most recent of these migrations, within the last 3,000 years, resulted in the colonization of the vast expanse of Polynesia.

Remarkably, it is among members of the Mormon church that we find some of the strongest resistance to mainstream views of New World and Pacific colonization. Not only do Mormons link Native American culture with ancient seafarers, most Latter-day Saints hold that the ancestors of indigenous Americans were Israelites, derived from small groups of immigrants who arrived hundreds of years before the birth of Christ. The Polynesians are believed to be descended from these maritime Hebrews as a consequence of further nautical excursions from their New World settlements. These beliefs are widely held among Mormons. For well over a century, such tenets have profoundly influenced church policies and played a major role in the conversion of indigenous peoples from both regions.

The staunch resistance to mainstream scientific views stems from Mormon faith in the Book of Mormon. First published in New York in 1830, it is believed by Mormons to be an American counterpart of the Bible describing the literal arrival and history of Hebrews in the New World. Joseph Smith, the prophet who brought forth the Book of Mormon, claimed the book was a direct translation from the record he said was inscribed on gold plates and buried in a hill near his home in the village of Manchester, New York. Smith said he went to the hill in 1827 and that the gold plates were delivered to him by an angel named Moroni, a prophet who had lived in America in about AD 400. According to Smith, Moroni was the last of a line of prophets who had written on the plates and the one who deposited them in what is now known as the Hill Cumorah. The Book of Mormon is considered by Mormons to contain a literal account of God's dealings with the people who lived anciently in the New World.

According to the Book of Mormon, most of its fifteen books were collated and abridged by the penultimate prophet-historian Mormon, after whom the book is named. Those who believe in the book's religious message, are, therefore, known as Mormons. The book is primarily devoted to a small group of Jews who, we are told, sailed from Jerusalem in 600 BC. The descendants of these colonists multiplied rapidly, splitting into two large nations. One nation is depicted as a culturally advanced society that was populated with a light-skinned race. The other nation was culturally inferior and was cursed by God with a dark skin. During most of the thousand-year Book of Mormon history, these light- and dark-skinned races remained in continual conflict. Eventually, the white-skinned nation descended into wickedness and was eliminated by the dark-skinned race around AD 400. It is to the descendants of the dark-skinned race that the Book of Mormon is most specifically addressed. Mormons believe that this race constitutes the principal ancestors of the American Indians.

The Book of Mormon is deeply embedded in the Mormon faith. Joseph Smith once affirmed that it was "the most correct of any book on earth," a claim that has been disputed since the day it was published. He went on to state that the book was "the keystone of our religion," which is undoubtedly true. The Book of Mormon was crucial to the establishment of the Mormon church. Adherents claim that if this record is true, then it follows that Joseph Smith was a true prophet of God. If Joseph Smith was a true prophet, then the Church of Jesus Christ of Latter-day Saints is the only true church on the face of the earth because Smith said so.

While its claims may appear extraordinary today, the Book of Mormon narrative mirrors the myths that permeated the society from which the church emerged. Most American colonists held to a very literal interpretation of the Bible, including the idea that there was a rapid colonization of the earth after the Flood in 2500 BC. The most widely accepted explanation for the origin of the so-called Red Man in the New World was that they were a degraded descendant of the scattered House of Israel. Indians were blamed for having annihilated another race that was believed to have been responsible for the construction of the elaborate buildings and cultural artifacts that American colonists uncovered as they advanced westward over the Appalachian Mountains. This other race was assumed to have been light-skinned.

To the dispassionate reader, the Book of Mormon is not the story of a small group that encountered an already largely populated America. The voyaging Israelites arrived in a land "kept from the knowledge of other nations" (2 Ne. 1:8). There is no mention of any non-Israelite people in the New World during the thousand-year period covered by the Book of Mormon. The narrative includes descriptions of large civilizations with populations reaching into the millions and the practice of Christianity, a written language, metallurgy, and the farming of several Old World domesticated plants and animals. In addition, the immigrant Hebrew Christians found horses, oxen, cattle, and goats in the New World.

Anthropologists and archaeologists, including some Mormons and former Mormons, have discovered little to support the existence of these civilizations. Over a period of 150 years, as scholars have seriously studied Native American cultures and prehistory, evidence of a Christian civilization in the Americas has eluded the specialists. In Mesoamerica, which is regarded by Mormon scholars to be the setting of the Book of Mormon narrative, research has uncovered cultures where the worship of multiple deities and human sacrifice were not uncommon. These cultures lack any trace of Hebrew or Egyptian writing, metallurgy, or the Old World domesticated animals and plants described in the Book of Mormon. Likewise in Polynesia, the accumulating scientific evidence suggests a west-to-east pattern of migration and a lack of any Old World cultural imprint before the arrival of white Europeans.

The absence of physical evidence supporting the Book of Mormon has had little impact on the millions of Mormons who consider the book to be a true record of the ancestors of Native Americans and Polynesians. Many LDS scholars have been eager to leap to the defense of the book and to criticize mainstream scientific views. The church employs academics at its own university who defend the Book of Mormon on a professional basis. Mormons are liberally provided with uplifting accounts of evidence that seems to support the book. Frequently this proof—and criticism of Gentile (non-Mormon) science—is delivered to church members by General Authorities speaking during world conferences. Consequently, Mormons remain deeply suspicious of Gentile theories, particularly any that conflict with widely accepted beliefs of the church.

However, the weight of evidence has forced Mormon scholars to rethink the scale, location, and nature of the historical account in the Book of Mormon. Over the past decade, there has been a marked shift among these scholars away from the views of the wider LDS community. Most LDS scholars today want to limit the Israelite colonization to the region of Mesoamerica, while a growing subset shrinks the book's claims even further. But seemingly oblivious to this revisionist scholarship, LDS leaders continue to teach that all or most Native Americans and Polynesians are literal descendants of the Israelites described in the Book of Mormon. The majority of faithful members believes likewise and resists the theories of LDS academics. Most Mormons of Native American or Polynesian ancestry—about one in five globally—believe that their family histories trace back to Israel.

The claim that Native Americans and Polynesians are the remnants of an early Diaspora is susceptible to investigation within a range of scientific disciplines, but it is the field of human genetics that provides Book of Mormon critics with the latest and most compelling evidence to challenge LDS claims. The recent sequencing of the human genome has captured the scientific spotlight. Less publicized has been the enormous progress in the field of human molecular genealogy, showing how our species emerged and spread across the earth. Human DNA genealogy reinforces the multi-disciplinary findings of how our ancestors spread throughout the earth over a period of many thousands of years to all the continents by at least 14,000 years ago. This research offers little comfort to those who are wedded to a literal interpretation of the Bible, which has our first parents walking the earth as recently as 6,000 years ago and all races springing from the loins of Noah a mere 4,400 years ago.

Molecular genealogists are now constructing DNA family trees of paternal and maternal ancestors and tracking the earliest human migrations around the world. These family trees have been particularly informative in such places as the New World and Polynesia, which are among the most recent areas colonized by humans. Molecular genealogy has allowed us to follow the footsteps of our ancestors, following the pathways of their genes, as they multiplied and replenished throughout every corner of the earth unto the isles of the sea.

An American Lost Tribe

1

A Chosen Race in a Promised Land

He called me by name, and said unto me that he was a messenger sent from the presence of God to me, and that his name was Moroni; . . . He said there was a book deposited, written upon gold plates, giving an account of the former inhabitants of this continent, and the source from whence they sprang.

—Joseph Smith, JS-History 1:33-34

Any attempt to describe Mormon doctrine is fraught with peril. The most exhaustive survey of the beliefs and doctrines of the church, the five-volume *Encyclopedia of Mormonism*, contains the proviso that its views are not necessarily the official positions of the church (Ludlow 1992). Yet, the encyclopedia was produced by church scholars, published jointly with church-owned Brigham Young University, and its production was supervised by General Authorities including a number of apostles. Over time, the Brethren have learned that sudden, public changes to long-held beliefs can be painful and damaging, as was apparent during the public reversals on polygamy and the ban on blacks holding the priesthood—doctrines that had become unsustainable in American society.

The most authoritative sources of LDS doctrine are the four "standard works." These include the King James Version of the Bible and the three uniquely Mormon scriptures, the Book of Mormon, the Doctrine and Covenants, and the Pearl of Great Price. Among Latter-day Saints, the power to "officially" interpret these scriptures resides with the prophets, seers, and revelators, or the First Presidency and the Quorum of Twelve Apostles. Recently the Brethren have emphasized that the words of the living prophets—the current First Presidency and apostles—take precedence over those of deceased prophets. Prior statements of belief are sometimes disregarded by current generations where they conflict with present-day teachings. Consequently, while most Mormon doctrine remains relatively stable, there are elements that remain fluid and changeable.

I will attempt to describe Mormon doctrines as they relate to the origins and racial characteristics of indigenous Americans. Most reference will be to Mormon scripture, particularly the Book of Mormon—the principal source of these uniquely Mormon beliefs. Other Mormon scriptures contain relevant revelations dictated by Joseph Smith in the years following publication of the Book of Mormon.

THE BOOK OF MORMON STORY

The Book of Mormon relates the story of the arrival of three groups of Middle Eastern refugees upon the American continent, the earliest in about 2200 BC and two later groups around 600 BC. The introduction to the current (1981) edition of the Book of Mormon reads:

> The Book of Mormon is a volume of holy scripture comparable to the Bible. It is a record of God's dealings with the ancient inhabitants of the Americas and contains, as does the Bible, the fullness [sic] of the everlasting gospel.
>
> The book was written by many ancient prophets by the spirit of prophecy and revelation. Their words, written on gold plates, were quoted and abridged by a prophet-historian named Mormon. The record gives an account of two great civilizations. One came from Jerusalem in 600 BC, and afterward separated into two nations, known as

> the Nephites and the Lamanites. The other came much earlier when the Lord confounded the tongues at the Tower of Babel. This group is known as the Jaredites. After thousands of years, all were destroyed except the Lamanites, and they are the principal ancestors of the American Indians.

The Book of Mormon is, for the most part, concerned with the thousand-year history of the descendants of Lehi, a prophet who, as the book claims, lived in Jerusalem immediately prior to the Babylonian captivity in about 600 BC (Figure 1.1). According to the Book of Mormon, Lehi's paternal lineage traced back via Manasseh to Joseph, the son of Israel, who was sold into Egypt (1 Ne. 5:14); however, Lehi is not mentioned in the Bible.

In the Book of Mormon, Lehi is warned by God to take his wife Sariah and family and flee into the wilderness (2:1-3) to avoid the impending destruction of Jerusalem. Lehi's family is joined by the family of Ishmael, also from Jerusalem, whose daughters become wives of Lehi's four sons, who are—beginning with the oldest—Laman, Lemuel, Sam, and Nephi. After an eight-year journey that takes them south near the Red Sea and then in an easterly direction, Lehi's party reaches the ocean at a place they name Bountiful (1 Ne. 17). The clarity with which this journey is described has led Mormons, scholars included, to believe that the group traveled south across the Arabian Peninsula to modern-day Yemen or Oman.

After arriving at Bountiful, Lehi's youngest and most righteous son, Nephi, is commanded by God to build a ship to carry his family across the sea. In spite of the resistance of Nephi's rebellious older brothers, Laman and Lemuel, they complete the task, and the small band of émigrés comprising approximately twenty individuals sets sail for the Promised Land. The route of the maritime voyage and the place of arrival in the New World are unspecified, but an easterly voyage skirting Australia is the most popular belief.

Upon arriving in the Americas, Lehi sees a vision of the land they had just left and casts a prophetic glance at their future in the land they have recently inherited:

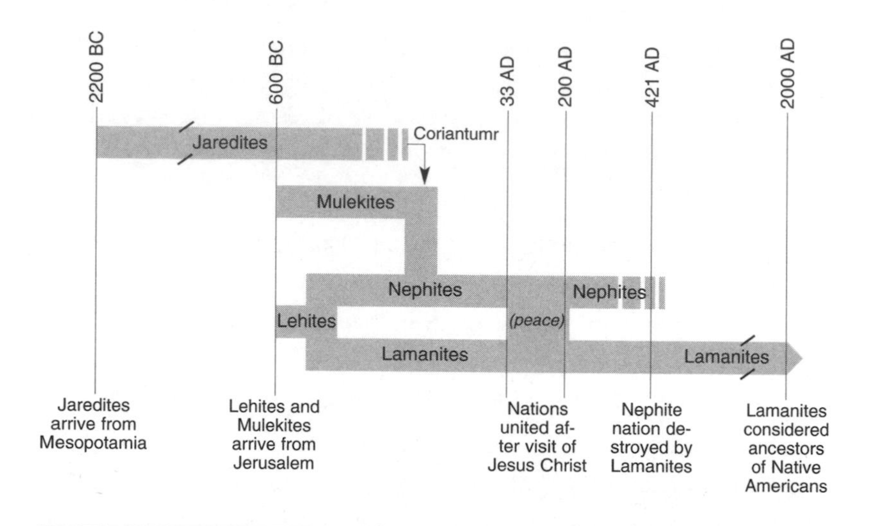

Figure 1.1 Basic chronology of New World civilizations described in the Book of Mormon

> For, behold ... I have seen a vision, in which I know that Jerusalem is destroyed; and had we remained in Jerusalem we should also have perished.
>
> But ... notwithstanding our afflictions, we have obtained a land of promise, a land which is choice above all other lands; a land which the Lord God hath covenanted with me should be a land for the inheritance of my seed. Yea, the Lord hath covenanted this land unto me, and to my children forever, and also all those who should be led out of other countries by the hand of the Lord.
>
> Wherefore, I, Lehi, prophesy according to the workings of the Spirit which is in me, that there shall none come into this land save they shall be brought by the hand of the Lord.
>
> And behold, it is wisdom that this land should be kept as yet from the knowledge of other nations; for behold, many nations would overrun the land, that there would be no place for an inheritance.
>
> Wherefore, I, Lehi, have obtained a promise, that inasmuch as those whom the Lord God shall bring out of the land of Jerusalem shall keep his commandments, they shall prosper upon the face of this land; and they shall be kept from all other nations, that they may possess this land unto themselves. (2 Ne. 1:4-6, 8, 9)

Two other groups undertake similar oceanic migrations to the Americas according to the Book of Mormon (Figure 1.1). The earliest are the Jaredites, who depart from Mesopotamia and the confusion of tongues at the Tower of Babel in about 2200 BC. The Jaredites are directed by God to travel into the wilderness to "that quarter where there never had man been" to a land of promise "preserved for a righteous people" (Ether 2:5-7). The Book of Mormon records that in fulfillment of prophecy, these people grow into a large civilization in the Americas with a population of several million. The Jaredites rapidly self-destruct in a series of cataclysmic battles between 279 and 130 BC, leaving a lone male survivor named Coriantumr.

Coriantumr is found by a third group of Middle Eastern émigrés, the Mulekites, who depart from Jerusalem at roughly the same time as Lehi. According to the Book of Mormon, a member of the founding party, Mulek, was a son of the biblical King Zedekiah who was carried off to Babylon. Despite the fact that the Lehites, Jaredites, and Mulekites undertake independent voyages to the Promised Land, all occupy neighboring territories in the New World. The Mulekites become assimilated into the Lehite nation and the Lehites find the ruins of the Jaredites in an area adjoining their civilization.

Soon after arriving in America, Lehi's party breaks into two opposing factions (Figure 1.1) based on sibling rivalry within Lehi's own family. The more righteous individuals follow Lehi's younger son Nephi, while the unbelievers follow his oldest son, Laman. The descendants of each group, from this time onward, are called Nephites and Lamanites, and they grow to become nations fated to be in almost continual conflict throughout a thousand-year period. The Lamanites become a scourge to the Nephites to "stir them up to the remembrance of their God" (2 Ne. 5:25).

Throughout the Book of Mormon account, frequent contrasts are drawn between the Nephites and Lamanites on the basis of relative cultural sophistication. The Nephites are said to be industrious and superior to the Lamanites in culture and technology. They raise herds of cattle, goats, and horses and till the earth, harvesting such Old World crops as wheat and barley. They are skilled craftsmen. They construct a temple similar in splendor to Solomon's. They are knowledgeable about iron, copper, brass, gold, and silver and find all of these precious metals in abundance in the Americas. They produce steel and fashion it into swords, breastplates, and arm and head shields to defend against the warring Lamanites. In response to frequent invasion, they build defensive mounds around their towns and cities. At the crest of these earthen ridges, they arrange wooden pickets and towers to overlook the fortifications.

In complete contrast to the Nephites, the Lamanites are idle and full of mischief, subtlety, and iniquity. They are lazy, idolatrous, and bloodthirsty. They dwell in tents and roam the wilderness hunting beasts of prey. They are skilled in the use of bows and arrows as well as slings. They shave their heads. The Nephites wear thick clothing into battle, but the Lamanites fight wearing only a short skin girdle. In spite of the fact that the Lamanites are backward, their civilization sustains populations that spawn armies vastly outnumbering the Nephites.

RACIAL DOCTRINE

In case the Nephites might be in any doubt about who the Lamanites are, God curses the latter "with a skin of blackness that they might not be enticing" to the Nephites who were "white, and exceedingly fair and delightsome" (5:21). If Nephites mix with the Lamanites, they risk a similar curse. On occasion, the curse is removed from the Lamanites when they are sufficiently righteous, but the curse returns when they become wicked again.

Book of Mormon descriptions of this curse represent a watershed in Mormon theology. To this day, they provide explicit scriptural statements linking the color of a person's skin to morality. Dark skin color was a deliberate mark applied by God as punishment for rebellious behavior. Early Mormons, weaned on a diet of nineteenth-century cultural chauvinism, felt comfortable with this starkly racist doctrine, as can be seen in this quote from an early LDS magazine:

> Is or is it not apparent from reason and analogy as drawn from a careful reading of the Scriptures, that God causes the saints, or people that fall away from his church to be cursed in time, with a black skin? Was or was not Cain, being marked, obliged to inherit the curse, he and his children, forever? ... Are or are not the Indians a sample of marking with blackness for rebellion against God's holy word and holy order? And can or can we not observe in the countenances of almost all nations, except the Gentile, a dark, sallow hue, which tells the sons of God, without a line of history, that they have fallen or changed from the original beauty and grace of father Adam? (*Latter-day Saints' Messenger and Advocate,* March 1835)

This insidious view of the superiority of the white race is clearly established in the Book of Mormon, which constitutes the earliest product of Joseph Smith's prophetic career. The church's connection with racial issues is well known in terms of its policies regarding Africans. But it was a subsequent revelation that entangled the church most deeply in the racial quagmire.

Genesis is short on details about the mark of Cain, so Joseph Smith, wanting to dispatch any ambiguity on the subject, produced a revision of Genesis—the canonized Book of Moses—into which a prophecy of Enoch was inserted explaining that the "seed of Cain were black" (Moses 7:22). In Mormon theology, all of humankind originated with Adam and Eve. In this respect, at least, the Genesis account is plainly understood. Later in the Bible, all of mankind except Noah's family are destroyed in the Flood, and Noah occupies an equally critical position of parenthood over the human family according to the Bible. Through Noah's sons, Shem, Ham, and Japheth, "was the whole earth overspread."

In a subsequent scripture dictated by Joseph Smith and known as the Book of Abraham—a book he said he translated from Egyptian papyri that had found their way into his hands—we learn how the mark of Cain survived the Flood. The lineage of Ham "preserved the curse in the land," the book reads; further that "Pharaoh, the eldest son of Egyptus, the daughter of Ham," settled in Egypt (Abr. 1:21-26). It is from this text that the awkward racial doctrine emerged, for Pharaoh was "of that lineage by which he could not have the right of the priesthood." Originally it was thought that the papyri from which Smith translated the Book of Abraham were destroyed in a Chicago fire. However, in 1966 the papyri were discovered in a museum in New York. Egyptologists, including a number of Latter-day Saints, have translated the texts and found them to be common funeral texts dating to about AD 100, the translations bearing no resemblance to the text of the Book of Abraham (Larson 1992).

Until 1978 the LDS Church denied the Mormon priesthood to Africans based on the authority of the Book of Abraham. Consequently, African men could not hold callings in the church or serve on missions. More critically, their families were denied certain temple rites that Mormons believe bind marriages and families together for eternity and are essential for salvation. Mormonism's institutionalized prejudice sparked some of the most severe criticism the church has faced. The theology reached rock bottom in an address given by Apostle Mark E. Petersen to LDS teachers of religion in 1954:

> Let us consider the great mercy of God for a moment. A Chinese, born in China with a dark skin and with all the handicaps of that race, seems to have little opportunity. But think of the mercy of God to Chinese people who are willing to accept the gospel. In spite of whatever they might have done in the pre-existence to justify being born over there as Chinamen, if they now, in this life, accept the gospel and live it the rest of their lives they can have the Priesthood, go to the temple and receive endowments and sealings, and that means they can have exaltation. Isn't the mercy of God marvelous?
>
> Think of the Negro, cursed as to the Priesthood. ... This negro [sic], who, in the pre-existence lived the type of life which justified the Lord in sending him to the earth in the lineage of Cain with a black skin, and possibly being born in darkest Africa—if that negro [sic] is willing when he hears the gospel to accept it, he may have many of the blessings of the gospel. In spite of all he did in the pre-existent life, the Lord is willing, if the Negro accepts the gospel with real, sincere faith, and is really converted, to give him the bless-

> ings of baptism and the gift of the Holy Ghost. If that Negro is faithful all his days, he can and will enter the celestial kingdom. He will go there as a servant, but he will get celestial glory. ...
>
> Now let's talk segregation again for a few moments. ... When the Lord chose the nations to which the spirits were to come, determining that some would be Japanese and some would be Chinese and some Negroes and some Americans, He engaged in an act of segregation. ...
>
> Who placed the Negroes originally in darkest Africa? Was it some man, or was it God? And when He placed them there, He segregated them. ... At least in the cases of the Lamanites and the Negroes we have the definite word of the Lord Himself that He placed a dark skin upon them as a curse ... [and] He forbade intermarriage ... [and] He certainly segregated the descendants of Cain when He cursed the Negro as to the Priesthood, and drew an absolute line. You may even say He dropped an Iron curtain there. ...
>
> We must not intermarry with the Negro, Why? If I were to marry a Negro woman and have children by her, my children would all be cursed as to the Priesthood. Do I want my children cursed as to the priesthood? If there is one drop of Negro blood in my children, as I have read to you, they receive the curse. There isn't any argument, therefore, as to inter-marriage with the Negro, is there? There are 50 million Negroes in the United States. If they were to achieve complete absorption with the white race, think what that would do. With 50 million Negroes inter-married with us, where would the Priesthood be? Who could hold it, in all America? Think what that would do to the work of the Church!
>
> Now we are generous with the negro [sic]. ... I would be willing to let every Negro drive a cadillac if they could afford it. I would be willing that they have all the advantages they can get out of life in the world. But let them enjoy these things among themselves. I think the Lord segregated the Negro and who is man to change that segregation? ... What God hath separated, let not man bring together again. (Petersen 1954)

In 1978 in response to considerable public pressure, the church announced a policy reversal and began allowing all worthy male members of the church to hold the priesthood. According to subsequent statements, there was considerable debate among the leaders when this adjustment was proposed for a doctrine that many General Authorities had believed would remain firmly in place until Christ returned to the earth. The flexibility with which Mormon theology was able to adapt to social change is noticeable in the rewriting of a few scriptural passages to tone down the references to skin color. The phrase "white and delightsome" found in the earliest edition of the Book of Mormon was changed in some cases (e.g., 2 Ne. 30:6) to "pure and delightsome," although not in all cases. The concept of skin color being a sign of God's displeasure is still evident (e.g. 1 Ne. 13:15; 2 Ne. 5:21; Jacob 3:8; Alma 3:6; 3 Ne. 2:15; Abr. 1:21-26). LDS scripture asserts that those who are "blessed" with a white skin are favored because of what they did as spirits in a pre-earth life.

FATE OF THE NEPHITES AND LAMANITES

About one-third of the Book of Mormon is devoted to a rather tedious procession of battles between the Lamanites and Nephites. The casualties arising from these conflicts provide frequent indications of the size of the displaced Hebrew populations. For example, in 190 BC a single battle claims the lives of 3,000 Lamanites (Mosiah 9:18). By 90 BC similar battles claim almost 20,000 lives (Alma 2:19). It is not uncommon for tens of thousands to be slain in a single year in the Book of Mormon. In addition, the book notes the departure of thousands of men, women, and children from the main centers of civilization into the "land northward." Some journey by ship, while others appear to travel on foot. By now, the population has multiplied until "they began to cover the face of the whole earth, from the sea south to the sea north, from the sea west to the sea east" (Hel. 3:8).

The Book of Mormon records that in AD 34 there is a break in hostilities brought about by the dramatic arrival of Jesus Christ soon after his death and resurrection. For almost two hundred years, the Nephites and Lamanites live in peace, practicing Christianity on the American continent. A short section, only twenty-three verses long, describes this astonishing period of harmony. After the lull, war resumes, and by about AD 330, armies numbering between 40,000 and 50,000 face each other (Morm. 2:9). Population growth has resulted in the whole face of the land being covered with buildings, and people have become "as numerous almost, as it were the sand of the sea" (1:7). During the last hundred years of their recorded history, these two nations pitch against each other in a seemingly irrational series of wars in which hundreds of thousands are slain. In the final battle, in approximately AD 385, a massive Lamanite army slaughters 230,000 Nephite men, women, and children (Morm. 6). The Lamanite population capable of sustaining an army of that size, capable of inflicting such carnage, must surely number in the millions.

From the time of the final Lamanite/Nephite conflict, as foretold by Lehi's son Nephi one thousand years earlier, the Lamanites remain in a degraded state, constantly at war among themselves:

> And it came to pass that I beheld, and saw the people of the seed of my brethren that they had overcome my seed; and they went forth in multitudes upon the face of the land.
>
> And I saw them gathered together in multitudes; and I saw wars and rumors of wars among them; and in wars and rumors of wars I saw many generations pass away.
>
> And the angel said unto me: Behold these shall dwindle in unbelief.
>
> And it came to pass that I beheld, after they had dwindled in unbelief they became a dark, and loathsome, and a filthy people, full of idleness and all manner of abominations. (1 Ne. 12:20-23)

The gold-plate record was written especially to the Lamanites, a remnant of the House of Israel, because it was known that they would survive the record-keeping Nephite prophets. A recurring message in the Book of Mormon is that the prophetic record would be brought to light in the latter days. Further prophecies predict the "scattering" and later "redemption" of the Lamanites by the Gentiles, who would restore to them the knowledge of their fathers. The Lamanites are apparently still in a degraded state when they clash with the Gentile Europeans who, it is known in the Book of Mormon, will arrive eleven centuries later:

> And now, I would prophesy somewhat more concerning the Jews and the Gentiles. For after the book of which I have spoken [Book of Mormon] shall come forth, and be written unto the Gentiles, and sealed up again unto the Lord, there shall be many which shall believe the words which are written; and they shall carry them forth unto the remnant of our seed.
>
> And then shall the remnant of our seed know concerning us, how that we came out from Jerusalem, and that they are descendants of the Jews.
>
> And the gospel of Jesus Christ shall be declared among them; wherefore, they shall be restored unto the knowledge of their fathers, and also to the knowledge of Jesus Christ, which was had among their fathers.
>
> And then shall they rejoice; for they shall know that it is a blessing unto them from the hand of God; and their scales of darkness shall begin to fall from their eyes; and many generations shall not pass away among them, save they shall be a pure and a delightsome people. (2 Ne. 30:3-6)

Numerous prophets in the Book of Mormon predict that the Lamanites are to be "driven and scattered by the Gentiles" and eventually gathered again (Hel. 15:12; Morm. 5:19-20; 3 Ne. 20:22; 21:23-26), this time to a New Jerusalem in America. Early volumes of the *Evening and Morning Star* and sermons recorded in *Discourses of the Prophet Joseph Smith* make frequent mention of the gathering being fulfilled right before the eyes of the church in the government's relocation of Indians to areas west of the Missouri River. It was in western Missouri in 1831 that Joseph Smith decided on the location for the city of New Jerusalem, a city to be built with the assistance of the Lamanites.

Mormons living in nineteenth-century North America felt a strong desire to bring to the Indians the blessings promised the Lamanites in the Book of Mormon. They saw this as a direct fulfillment of ancient prophecies, and Joseph Smith reiterated this theme in several of the earliest revelations that he dictated, now contained in the Doctrine and Covenants:

> And this testimony shall come to the knowledge of the Lamanites, and the Lemuelites, and the Ishmaelites, who dwindled in unbelief because of the iniquity of their fathers, whom the Lord has suffered to destroy their brethren the Nephites, because of their iniquities and their abominations.
>
> And for this very purpose are these plates preserved, which contain these records—that the promises of the Lord might be fulfilled, which he made to his people;
>
> And that the Lamanites might come to the knowledge of their fathers, and that they might know the promises of the Lord, and that they may believe the gospel and rely upon the merits of Jesus Christ, and be glorified through faith in his name, and that through their repentance they might be saved. (3:18-20)

The restoration of the Lamanites was thought to involve much more than providing the Indians with knowledge of their fathers and the Christian god. The Lamanites were to be returned to the civilized state from which they had degenerated, assisted by the very Gentiles who were simultaneously appropriating their lands.

2

Race Relations in Colonial America

... the belief that Europeans and Native Americans are at different stages of development has underpinned European attitudes since the time of Columbus. Through the centuries, it has validated the certainty that some force greater than ourselves (God, history, evolution) destines Europeans and Euro-Americans—for better or worse—to subdue the wilderness and supplant the "Indian." ... From our Olympian perspective at the pinnacle of creation, there can be no permanent co-existence, no equality, between the "objective" reality we see and the legends of more "primitive" people.

If we are to begin to understand the experience of Native Americans we have to challenge the tyranny which this view has established in our minds.

—James Wilson, 1999

Many scholars have looked beyond ancient American prophets, angelic visitors, and gold plates to the societal milieu from which the Book of Mormon emerged to deduce the book's genesis. A palpable similarity exists between descriptions of the degraded Lamanite race and Indian stereotypes that were widely accepted in the community in which Joseph Smith lived (Brodie 1971). An exploration of the prejudices of Joseph Smith's era and the historical factors that contributed to their development helps to clarify how aspects of America's northeastern frontier culture found its way into the Book of Mormon. Fortunately, some present-day understandings of the pre-settlement native world have been liberated from the prevailing opinion of earlier centuries (McNickle 1971). A more complete review of the environmental influences on Joseph Smith can be found in Dan Vogel's *Indian Origins and the Book of Mormon.*

EUROPEAN CONQUEST

Permanent European settlement of North America first occurred on the Atlantic coast, predominantly amid the Algonquian tribes who inhabited most of the coastal regions (King 1999). Iroquoian-speaking nations dominated what is now upper New York State and the lands adjoining Lake Ontario and the St. Lawrence Valley. Native societies depended on a seasonal balance of hunting and gathering. In addition, men typically cleared fields for cultivation and hunted for deer, turkey, and fish while women carried out the bulk of the farming activities associated with the cultivation of maize, beans, and squash (Trigger 1978). In some cases the survival and prosperity of early white settlements relied on the assistance of local Native American tribes who taught the newcomers how to cultivate corn and how to fish, hunt, and exploit local flora and fauna.

Colonies of European settlers on the Atlantic seaboard expanded rapidly due to high fecundity and a steady stream of immigrants. Native Americans coveted European cloth and metal goods, while New England colonists appropriated the most fertile agricultural lands, frequently under dubious circumstances (King 1999). The plight of the coastal tribes became steadily more desperate throughout the seventeenth and into the eighteenth century as they suffered the ravages of disease and land deprivation and were increasingly squeezed between rival European settlements. The Algonquian people, who bore the brunt of colonization, largely faded from history. The few survivors migrated to the west over the Appalachian Mountains into the Ohio Valley.

When Europeans arrived, the Indians immediately west of the Appalachian Mountains already lived in permanent settlements throughout the Mississippi and Ohio Valleys (Hunter 1978). However, the shock waves of early contacts were felt on the well-trodden trade and communication networks, penetrating deep into the interior of North America. Waves of highly infectious smallpox, whooping cough, and other infections spread rapidly along these thoroughfares, usually preceding the first physical contact between natives and Europeans (Trigger 1978). Native Americans had virtually no immunity to these diseases and native populations were decimated; disease frequently eliminated whole communities.

By the time the first white settlers arrived in the Mississippi and Ohio Valleys, Native Americans living there were the remnants of a number of shattered tribal nations, many of whom had been evicted from their homelands. They included coastal tribes such as the Shawnee, as well as the Delaware and the Seneca (Iroquois) from what is now upper New York State. Other tribes drawn to the valley included the Miami and Wyandot from regions near Lake Michigan (Trigger 1978). In the late eighteenth and early nineteenth centuries, large numbers of Europeans streamed into the Ohio Valley in search of fertile farming land. Increasingly, native populations found themselves caught up in a series of conflicts between the French, British, and American colonists, eventually surrendering their claim to Ohio lands through a series of treaties with the United States. With the Louisiana Purchase of 1803, President Thomas Jefferson acquired additional land west of the Mississippi River. In 1818 the government, under President Andrew Jackson, began the forced removal of many of America's native outcasts farther west to the new frontiers. Southeastern native tribes were forced into Oklahoma Territory (Jennings 1993).

EUROPEAN VIEWS OF NATIVE AMERICANS

All three major European civilizations vying for a piece of the New World saw Native Americans through the "civilized" eyes of Christian cultures. The early Spanish conception was quickly adopted by French and English colonists, who shared a comparable religious and cultural heritage. Greater knowledge of other cultures on other continents served to increase European self-assurance (Berkhofer 1978), seeing Europe in terms of intellectual, cultural, military, and political superiority. Comparisons between European and Native American civilizations formed the backdrop of descriptions and understandings of indigenous people from the very beginning of the white conquest of America. European colonizers often concluded that the contemporary Indians had made little impact on North America and had therefore squandered some of the most productive land known to man.

To invading Europeans, Christian civilization stood unchallenged at the top of the cultural ladder. Savage, barbarous, and darker-skinned "tribes" occupied the lower rungs. The semi-nomadic course of life and perceived lack of civil order of Native Americans were thought to be evidence of a lack of civility (Axtell 1981). Natives were presumed to be enslaved to passion and a race of idlers. Hunting and fishing, the pursuits undertaken by native men, were viewed as recreation. In England, hunting was not considered an important economic activity but a pastime of aristocrats and shrewd poachers. Europeans were critical of native farming methods which involved the use of crude hoes and handmade baskets. Yet, Native Americans had, by these means, achieved sufficient yields for their needs.

From the earliest days of entry into the New World, French and English colonizers had attempted to spread Christianity into native realms to save the soul of the savage. The Judeo-Christian worldview portrayed mankind as detached from the natural world and an outcast from the presence of the Creator, charged with the responsibility of subduing the earth. Through God's written word, Europeans had little opportunity to learn about how things were in the greater world, but assumed the Bible's plenary authority to judge other people's beliefs as factually wrong (Axtell 1981). In the New World, Europeans were suddenly confronted with unknown people who were not mentioned in scripture. Genesis needed to be reconciled with the existence of Native Americans, and Native Americans reconciled with Genesis and the repopulation of the earth after the flood of Noah as recently as 2500 B.C. Native Americans had no record of these landmarks in the earth's history, nor did they remember them.

When American Indians came to be accepted as part of the human race, the problem became one of trying to link them into the family tree that sprouted from Adam and Eve. On the basis of superficial comparisons of language, dress, religious rites, and cultural traits, various observers linked Native American cultures with almost every culture known to man (Berkhofer 1978). These ranged from ancient Greeks, Scythians, Tartars, and Hebrews to the Spaniards themselves, as well as Danes, Welsh, and even the people of Atlantis. The ancient Hebrews were a popular choice, usually in the hope of providing a connection between the American Indian and the Ten Lost Tribes. Some thinkers on the subject favored a theory that was first proposed in 1590 by José de Acosta in his *Natural and Moral History of the Indies*, that Native American cultures arose separately from the Old World and that the original inhabitants had migrated over a land bridge from Asia.

A growing awareness of the extent of cultural diversity in the Americas made it difficult to reconcile this diversity with the biblical chronology. Traditionally racial diversity had been accounted for by the separate post-Flood migrations of Noah's sons and the variety of languages that emerged after the destruction of the Tower of Babel. But how had such striking diversity arisen in America in such a short time frame and at such a long distance from the cradle of civilization in the Middle East? A dominant line of reasoning gradually emerged that Native American cultures must have been corrupted forms of higher Old World civilizations. Well into the nineteenth century, scholars and non-scholars alike evaluated Native Americans on the basis of a degenerate form of white culture, an impression that helped rationalize the unrelenting expansion of European colonization.

Europeans also saw native people and their cultures as a single entity. They applied the label "Indian" broadly to cultures with a plurality of customs and wide diversity of values and beliefs (Berkhofer 1978). Native Americans lived in at least two thousand different societies, reflecting the wide geographic and climatic range found in the New World. In North America alone, there were about 600 separate populations occupying eleven distinct cultural areas (Wilson 1999). Even the term Indian was born of misunderstanding, when Columbus named the inhabitants of the Bahamas Indians, thinking he had reached one of the myriad, nameless islands in the Sea of Indies below Asia and unaware that a new continent had interrupted his journey (Lyon and Sacha 1992). The image of the Indians portrayed by Europeans was a stereotype that bore little similarity to how the original inhabitants of the western hemisphere had seen themselves at the time of the first European contact.

Clearly, there are striking similarities between the Lamanite race as a naked, head shaven, tent dwelling, bow and arrow wielding, idle, thieving, bloodthirsty hoard and the deeply entrenched Native American stereotypes of the period. The Book of Mormon account transports us directly back to the moment in America's history when zealous patriotism, speculation, and racism were rampant. At this juncture, the colonial thrust of Europeans generated a period of heightened misunderstanding between Native and European Americans. It is probable that Joseph Smith wove these frontier prejudices into the Book of Mormon without question, unaware that the common knowledge of his time would be subject to considerable historical shifts and revision in the years to come.

But the misunderstanding of indigenous people and their cultures would descend to even greater depths. From the ferment of colonial gossip emerged the myth that was to profoundly impact relations between white settlers and indigenous Americans for a century. Speculation arose that Native Americans had committed ancient atrocities infinitely greater than any injustice they had experienced at the hands of colonial Americans. In the darkest chapter in their history, the Indians found themselves accused of having carried out an extraordinary act of genocide, the complete annihilation of an enlightened, fair-skinned race that was thought to have occupied the Americas hundreds of years before the arrival of the Europeans.

THE MOUND BUILDER MYTH

In 1772 a group of Christian Indians led by David Zeisberger discovered curious earthworks at a settlement they founded near the present-day site of New Philadelphia, west of the Ohio River. As the land in the adjacent valleys was cleared, thousands of mounds were uncovered which had lain hidden under centuries of forest regeneration. In isolation, few of the mounds were impressive structures, but the sheer number of them had a remarkable effect on the new colonists. In the Ohio Valley alone, the center of mound building activity, as many as 10,000 mounds were identified. Hardly a valley was colonized where mounds were not found, many of which were leveled to make way for farmland. The mounds captured the imagination of the settlers. Grave digging for artifacts, especially in the more elaborately adorned mounds, became a popular pastime (Silverberg 1968).

The artifacts—decorated pipes, jewelry, breastplates, and ornaments—were crafted from native copper and occasionally coated in gold and silver, clearly indicating that there had been a nation of skillful artisans. To the European settlers for whom the Indians of the Ohio Valley did not appear capable of building the mounds or possessing the metallurgical skills necessary to produce the copper, gold, and silver artifacts, the mounds remained a mystery. Nor did the displaced Ohio Indians have a tradition that would explain the construction of the mounds. The fact that two centuries earlier the explorer De Soto observed contemporary Indians engaged in mound building activity further south was not widely known. Consequently, the mounds were attributed to an advanced race that had been swept from the face of the American continent. The foundation was in place for the elaboration of the Mound Builder myth, spun by Europeans without any substantiation from native peoples (Silverberg 1968).

It was soon widely held that a civilized race had come from the Old World and built the mounds. The impressive size and number of the structures confirmed the skill of the builders, and it seemed that the populations required to sustain such artisans would have been greater than one would find among a hunting and gathering culture. An early survey of the antiquities of Ohio by Caleb Atwater in 1820 concluded that the tumuli "owe their origin to a people far more civilized than the Indian but far less so than Europeans." Compelling evidence for the cultural superiority of the Mound Builders was their alleged level of skill in metallurgy. Atwater described artifacts of pure copper and what he interpreted to be corroded remains of a steel sword and cast iron plates. It would be decades before it was discovered that the copper originated from unusually pure natural deposits in Michigan, that the gold and silver had originated in alluvial deposits, and that the artifacts which appeared to be made of steel were likely made of meteoric iron. However, Atwater's observations seemed to confirm the conventional wisdom about the Mound Builders having been skilled in the mining and smelting of iron, copper, gold, and silver.

Frontier antiquarians speculated about the origins of the Mound Builders. Alternative explanations included theories about Vikings who may have stopped off in the Ohio and Mississippi Valleys on their way to colonize Mexico to build even more magnificent pyramids and temples. An opposing theory was that Mexican colonists had migrated into the Mississippi Valley, gradually forgetting their mound-building skills as they moved north. This explained the declining complexity of the mounds at the northern extremes of the Ohio cultures. Others contrived tales of heroic voyages of Phoenicians, Greeks, or Welsh explorers. The most persistent theory was that the mounds were built by descendants of the Lost Tribes of Israel who had made their way across the ocean after their dispersal by the Assyrians (Silverberg 1968).

For those who subscribed to the Lost Tribe theory, the cause of the disappearance of the enlightened race and discontinuance of earthworks was simply attributed to the despised local Indians. The remains seemed to give clues to a fierce struggle between the enlightened Mound Builders and a bloodthirsty heathen nation. Some of the mounds had defensive timber palisades, evoking visions of desperate attempts to fight off the barbarian hoards. The presence of thousands of mounds containing human remains was evidence of the vast numbers killed and the intensity and frenetic pace with which the conflict must have taken place. Few antiquarians at the time would have known that the local Indians customarily exhumed, collected together, and reburied the bones of the deceased in a ceremony known as the Festival of the Dead. The Indian forts atop some of the mounds were, in many cases, the handiwork of the contemporary Iroquois tribes. Implicating the Indians in the annihilation of a superior race with the trappings of advanced European civilizations further served to alienate Native Americans:

> In such an intellectual environment it was impossible for the conservatives to make themselves heard and almost impossible for them to find a following. Some deep national need was fulfilled by the myth of the Mound Builders, and debunkers were unpopular. The dream of a lost prehistoric race in the American heartland was profoundly satisfying; and if the vanished ones had been giants, or white men, or Danes or Toltecs, or giant white Jewish Toltec Vikings, so much the better. The people of the United States were then engaged in undeclared war against the Indians who blocked their path to expansion, transporting, imprisoning, or simply massacring them; and as this century-long campaign of genocide proceeded, it may have been expedient to conjure up a previous race whom the Indians had displaced in the same way. Conscience might ache a bit over the uprooting of the Indians, but not if it could be shown that the Indians, far from being long established settlers in the land, were themselves mere intruders who had wantonly shattered the glorious Mound Builder civilization of old. What had been a simple war of conquest against the Indian now could be construed as a war of vengeance on behalf of that great and martyred ancient culture. (Silverberg 1968, 57-8)

Debate over the Mound Builders raged through most of the nineteenth century. Amateur antiquarians continued to excavate the mounds, and numerous popular books on either side of the Atlantic fueled the myth with speculations about where the Mound Builders had come from, when they had prospered, and where they had gone (Willey 1980). In the haste of colonization, most of the mounds were eventually leveled and their contents plundered. For a time, it appeared that nothing would remain and that the secrets of the mounds would be lost forever; however, in the latter half of the nineteenth century, less speculative studies were undertaken. The Smithsonian Institution, founded in 1846, and Harvard University's Peabody Museum, founded in 1866, spurred the emergence of professional standards for archaeology and a more descriptive and methodical approach. An important contribution was the work of Ephraim Squier and Edwin Davis, *Ancient Monuments of the Mississippi Valley,* published in 1848 by the Smithsonian Institution. Increasingly, scholars opposed the prevailing Mound Builder hypothesis, arguing instead that the earthworks contained the deceased ancestors of the existing Native Americans. The myth was finally demolished in 1894 by Cyrus Thomas of the Bureau of Ethnology, a research arm of the Smithsonian Institution. Thomas published a monumental work that presented all of the bureau's data on the mounds, essentially ending an era of speculation among professional archaeologists (Willey 1980). The large burial mounds, elaborate mortuary practices, and geometric earthworks of the Ohio and Illinois Valleys were the handiwork of Native Americans who practiced what are now known as the Adena and Hopewell cultural traditions. Adena mortuary practices reached back 1,800-2,500 years ago in the central Ohio Valley. Hopewell traditions developed slightly later, within a period from 1,600-2,100 years ago, from centers in the Ohio and Illinois Valleys. However, the Mound Builder myth, which had dominated white American thought for a century, would influence white perceptions of Native Americans for many decades beyond.

FRONTIER SCRIPTURE

Joseph Smith was raised in western New York at the northern periphery of the major mound building centers. The whole region was rich in Indian relics, and hundreds of mounds dotted the countryside. The Smith home, near the town of Palmyra in Ontario County (now Wayne County), was located within a few miles of at least eight mounds. There was a contagious excitement over the possibility of finding Indian treasure, and treasure hunting became a minor obsession for many. The Smith family, in particular Joseph, fell under the spell of the mounds and could not resist the lure of buried riches. Residents in eighteen locations around Manchester, South Bainbridge, Colesville, and Windsor in New York State and Harmony, Pennsylvania, were witnesses to Smith family treasure quests (Vogel 1986, 1994). When Joseph Smith was twenty years old, he admitted using a "seer" or "peep" stone to assist him "locate hidden treasures in the bowels of the earth." This extraordinary confession was elicited during a trial in Bainbridge in March 1826 when he had been charged with being disorderly and an impostor in consequence of his "money digging" activities. Eighteen months later, Joseph Smith came into possession of the gold plates from which the Book of Mormon was translated.

By the time the Smith family arrived in Palmyra, most of the local Indians had been relocated to reservations, but the Mound Builder myth was widely accepted and the antiquities of the region further served to keep the myth alive. Palisaded forts and tumuli dotted the landscape from which skillfully wrought copper ornaments were uncovered. Numerous skeletons were uncovered in some of the mounds, leading some to regard them as mass graves. Palmyra newspapers frequently contained articles carrying the familiar conjecture surrounding the Mound Builders (Brodie 1971), and it was widely speculated that a great slaughter of a more civilized race had taken place in the surrounding area.

The longing of many colonial clergymen to link the Indians with the Lost Tribes of Israel sparked a proliferation of books outlining outrageous parallels between Indian and Hebraic cultures. Typically the authors of these books selectively sifted the growing anecdotal data for whatever similarities could be imagined to support this popular belief. Elias Boudinot's book entitled *A Star in the West; or a Humble Attempt to Discover the Long Lost Tribes of Israel in America Preparatory to Their Return to Their Beloved Jerusalem* (1816) epitomizes the genre. The epic contains countless pages itemizing parallels between Indian and Hebrew cultures, preceded by an extensive analysis of the Bible showing that the migration to America was in fulfillment of prophecy. Other examples of this type of colonial speculation include James Adair's *The History of the American Indians* (1775), Charles Crawford's *Essay upon the Propagation of the Gospel, in which there are facts to prove that many of the Indians in America are descended from the Ten Tribes* (1799), and Josiah Priest's *The Wonders of nature and Providence Displayed* (1825).

Scholars have concluded that the inspiration for the Hebrew origins of the migrants described in Joseph Smith's Book of Mormon came partly from the book *View of the Hebrews; or the Ten Tribes of Israel in America* (Persuitte 2000). This popular book was published in 1823, with a second edition in 1825, by Ethan Smith (no relation), pastor of a church in Poultney, Vermont. In Ethan Smith's book, a New World history is contemplated that shares close parallels with the plot of the Book of Mormon:

> The probability then is this; that the ten tribes, arriving in this continent with some knowledge of the acts of civilized life; finding themselves in a vast wilderness filled with the best game, inviting them to chase; most of them fell into a wandering idle hunting life. Different clans parted from each other, lost each other, and formed separate tribes. Most of them formed a habit of this idle mode of living, and were pleased with it. More sensible parts of this people associated together, to improve their knowledge of the arts; and probably continued thus for ages. From these the noted relics of civilization discovered in the west and south were furnished. But the savage tribes prevailed; and in process of time their savage jealousies and rage anni-

> hilated their more civilized brethren. And thus, as a holy vindictive Providence would have it, and according to ancient denunciations, all were left in an "outcast" savage state. This accounts for their loss of the knowledge of letters, of the art of navigation, and of use of iron. ... It is highly probable that the more civilized part of the tribes of Israel, after they settled in America, became wholly separated from the hunting and savage tribes of their brethren; ... [and] that tremendous wars were frequent between them and their savage brethren, till the former became extinct. (*View of the Hebrews* 1825, 172-73)

Further similarities between the two books include extensive quotations of the prophecies of Isaiah and the portrayal of a role for America in the last days for gathering the remnants of the House of Israel. Both authors paid particular attention to a prophecy of Ezekiel concerning the House of Israel:

> Moreover, thou son of man, take thee one stick, and write upon it, For Judah, and for the children of Israel his companions: then take another stick, and write upon it, For Joseph, the stick of Ephraim, and for all the house of Israel his companions:
>
> And join them one to another into one stick; and they shall be one stick; and they shall become one in thine hand. (Ezek. 37:16-17)

Joseph Smith believed that the Book of Mormon directly fulfilled this prophecy, and he promoted his new book as "the stick of Joseph taken from the hand of Ephraim" (Brodie 1971).

In spite of its extensive similarities with the Book of Mormon, *View of the Hebrews* should not be regarded as the sole source of inspiration for the book. The basic themes running through both publications merely reflected the most commonly accepted myths surrounding the mounds, the Indians, and the original colonization of America. The principal difference is that Ethan Smith's work was open speculation, whereas the Book of Mormon was a narrative that purported to be a literal, eyewitness account of what happened.

While the Mound Builder myth was widely accepted in the early nineteenth century, scholars were already beginning to wonder how long the existing Native American cultures may have been resident in the Americas. Joseph Smith likely noted the speculation about whether Indians might have migrated to the Western Hemisphere after the Flood in light of the growing awareness of the enormous diversity of Indian cultures and languages. Another problem Smith is likely to have grappled with was how animals arrived in America after the Flood. Within the concluding pages of the Book of Mormon, we find the brief account of the Jaredites, who are said to have sailed to the Americas at the time of the Tower of Babel. Joseph Smith went to the trouble of pointing out that the Jaredites brought to America "their flocks which they had gathered together, male and female of every kind." The Jaredites also brought "fowls of the air ... fish of the waters ... and seeds of every kind" (Ether 2:1-3). Smith became convinced that any earlier arrivals were out of the question. The Jaredites had arrived in a land "preserved for a righteous people," where mankind had never been and where the settlers were free "from captivity ... from all other nations under heaven." About 2,000 years after their arrival, the Jaredites "ripened in iniquity" and were swept from the face of the land, leaving it in an uncluttered state for the family of Lehi to inhabit.

The white man's perceptions of Native Americans and the Mound Builder myth, both of which permeated the New England society of Joseph Smith's day, became embedded in Mormon scripture. In many respects, the characteristics of the Book of Mormon Lamanites mirror the misunderstandings that surfaced in the froth of frontier speculation. The Mound Builder myth receives scriptural confirmation in the closing chapters of the Book of Mormon story where the final destruction of the fair-skinned, civilized Nephites occurs at the hand of their brethren, the savage, dark-skinned Lamanites. The story must have appeared plausible to early Americans who, for most of the nineteenth century, believed that Native Americans were responsible for the genocide of the postulated earlier, advanced race. The stereotypes and misunderstandings served to validate the Europeans' theft of native lands as an act of retribution; American Indians were themselves intruders in a land that had belonged to an earlier race—one that was comfortingly familiar to white colonists.

These stereotypes ignored the plurality of native cultures that existed in the New World before contact with Europeans. The myth relied heavily on value judgments of native cultures that clearly placed "savage" races lower on the ladder of civilizations and respectability. And the prejudices reflected a judgment passed on native cultures at the lowest point in their existence, at a time when the very framework of Native American societies had been torn apart by disease, war, and by an irrepressible invading force that was steadily taking possession of their homelands.

3

Lamanites in the Latter Days

But before the great day of the Lord shall come, Jacob shall flourish in the wilderness, and the Lamanites shall blossom as the rose.

—D & C 49:24

EARLY PERCEPTIONS

From the moment the Book of Mormon was published, Mormons believed it to contain the literal history of the ancestors of the American Indians. The Indians were thoutht to be the degraded descendants of Lehi, the Lamanites, who had wiped out the righteous, white-skinned Nephites. The similarities between the Lamanites and the despised Red Man were apparent, convincing many early settlers of the authenticity of the Book of Mormon. These beliefs are starkly revealed in the writings of church leaders in early church periodicals:

> The Nephites who were once enlightened, had fallen from a more elevated standing as to favor and privilege before the Lord, in consequence of the righteousness of their fathers, and now falling below, for such was actually the case, were suffered to be overcome, and the land was left to the possession of the red men, who were without intelligence, only in the affairs of their wars; and having no records,

> only preserving their history by tradition from father to son, lost the account of their true origin, and wandered from river to river, from hill to hill, from mountain to mountain, and from sea to sea, till the land was again peopled, in a measure, by a rude, wild, revengeful, warlike and barbarous race. Such are our Indians. (*Latter-day Saints' Messenger and Advocate,* July 1835)

Joseph Smith believed that native North Americans were a remnant of the Lamanites. The terms Indian and Lamanite were used interchangeably in his personal writings, many of which are contained in the *History of the Church of Jesus Christ of Latter-day Saints* compiled by B. H. Roberts (Smith 1932). Smith boasted that the Book of Mormon solved the mystery of the origins of America's ancient inhabitants, choosing to be equally explicit with members and non-members alike. When John Wentworth, editor of the *Chicago Democrat,* enquired about the rise of Mormonism, Smith included in his reply a candid outline of the racial history of the Americas as revealed within the pages of the Book of Mormon (Jessee 1984):

> In this important and interesting book the history of ancient America is unfolded, from its first settlement by a colony that came from the tower of Babel, at the confusion of languages to the beginning of the fifth century of the Christian era. We are informed by these records that America in ancient times has been inhabited by two distinct races of people. The first were called Jaredites and came directly from the tower of Babel. The second race came directly from the city of Jerusalem, about six hundred years before Christ. They were principally Israelites, of the descendants of Joseph. The Jaredites were destroyed about the time that the Israelites came from Jerusalem, who succeeded them in the inheritance of the country. The principal nation of the second race fell into battle towards the close of the fourth century. The remnant are the Indians that now inhabit this country. (Joseph Smith 1842)

Between 1830 and 1831, Smith dictated several revelations in which early elders of the church were instructed by God to embark on missions to the Lamanites. They were told to visit them on the Missouri frontier—the "borders of the Lamanites." At this time in America's history, Native Americans were being relocated to "Indian Territory" at the western edge of Iowa and Missouri. Several church elders, including Parley Pratt and Oliver Cowdery, preached to Native Americans in those states, and Mormons saw the repatriation of Indians to the western extremities of colonial America as a direct fulfillment of prophecy:

> It is not only gratifying, but almost marvellous [sic], to witness the gathering of the Indians. The work has been going on for some time, and these remnants of Joseph gather by hundreds and settle west of the Missouri, and Arkansas. And is not this scripture fulfilling: Give ear, O Shepherd of Israel, thou that leadest Joseph like a flock, through the instrumentality of the government of the United States? For it is written, Behold I will lift up my hand to the Gentiles, and set up my standard to the people: and they shall bring thy sons in their arms, and thy daughters shall be carried upon their shoulders. Thus said the prophet and so it is ... (*Evening and Morning Star*, Dec. 1832)

Relations between Mormons and indigenous Americans were limited to missionary ventures to Indian reservations. The church leadership never seriously questioned the propriety of the repatriation of Indians from their homelands to the reservations. Like most other colonists, Mormons saw themselves as the rightful owners of the most productive land in the United States. This remarkable quote, borrowed from fellow Gentile colonists, appeared in the *Evening and Morning Star* two years after publication of the Book of Mormon. It was headlined "MAN was created to dress the earth, and to cultivate his mind, and glorify God":

> The following is found in an ancient History of Connecticut. Soon after the settlement of New-Haven; several persons went over to what is now the town of Milford, where, finding the soil very good, they were desirous to effect a settlement; but the premises were in the peaceable possession of the Indians, and some conscientious scruples

> arose as to the propriety of deposing and expelling them. To test the case a Church meeting was called, and the matter determined by the solemn vote of that sacred body. After several speeches had been made in relation to the subject, they proceeded to pass votes—the first was the following;—Voted, that the earth is the Lord's and the fullness thereof. This passed in the affirmative, and, "Voted, that the earth is given to the saints."—This was also determined like the former … 3d. "Voted, that we are the saints," which passed without a dissenting voice, the title was considered indisputable, and the Indians were soon compelled to evacuate the place and relinquish the possession to the rightful owners. (*Evening and Morning Star,* June 1832)

Similar scruples surfaced when Mormons arrived in the Great Salt Lake Valley in 1847. They found themselves uneasy neighbors with several Native American tribes that had learned to survive in the harsh desert environment. The immigrants soon occupied most of the arable land and fresh water resources in Utah, leaving the Native Americans few options. The expansion of Mormon settlements with their accompanying European diseases and the settlers' zealous preoccupation with building Zion nullified any hope of a fair coexistence between the two groups. Conflict was inevitable and only abated when Brigham Young enlisted the support of the United States government to force the last remaining Native Americans onto reservations (Chadwick and Garrow 1992).

Few Native Americans became assimilated into the Mormon culture until the 1940s when leaders began to reemphasize their saintly responsibility toward the Lamanites. Indian missions were established in the United States, in Central and South America, and in Polynesia. The church established the Indian Placement Program which placed Native American children into white foster families to improve their educational standards. Between 1954 and 1996, about 70,000 native youth passed through the program. While the renewed focus on Native Americans through Lamanite missions and the Indian Placement Program brought improved health, education, farming techniques, and community development, important motives behind the establishment of the program were to preach the gospel and civilize the Indians.

CONTEMPORARY BELIEFS

The history and prophecies of the Book of Mormon are intertwined with the history of America, and the book continues to shape LDS understanding of Native Americans. Mormonism sees little value in native cultures and their ritualized histories and mythology; "Indian" cultures share such uneasy similarities to negative Book of Mormon images of the degraded Lamanites. Instead, Mormonism seeks to displace these to fulfill the prophetic destiny of the Lamanite.

One hundred and seventy years after its publication, the Book of Mormon still holds center stage in the unfolding drama of Mormonism. As a direct consequence of this book, most Native American Latter-day Saints throughout the Americas regard the Israelite Lehi to be a blood relative. In sermons, prayers, magazines, lesson manuals, and books, leaders have repeatedly spoken of the Lamanite birthright of native peoples. With full prophetic support, the modern Lamanite family has expanded to include not only Native Americans but also the Polynesians:

> The Lord said that when his coming was near, the Lamanites would become a righteous and respected people. He said, "Before the great day of the Lord shall come, ... the Lamanites shall blossom as the rose" (D&C 49:24). Great numbers of Lamanites in North and South America and the South Pacific are now receiving the blessings of the gospel. (*Gospel Principles* 1997, 268)

The emergence of Lamanite folklore among the Polynesians deserves particular attention and is the subject of the next chapter.

Among the most authoritative vehicles for reinforcing genealogical bonds with Father Lehi are institutionalized blessings, known as patriarchal blessings, that are bestowed upon adult members of the LDS church. Patriarchal blessings are considered direct revelations to the recipients from God. They are called patriarchal because they disclose a recipient's patriarchal Israelite lineage. Israel lies at the center of Mormon theology. Like Jews, Mormons believe that they are God's chosen people, adopted into the ecclesiastical kingdom of Israel upon baptism into the church. The patriarchal blessing generally promises good fortune for the recipient, then reveals which tribe of the House of Israel the person belongs to, either through adoption or literal blood ties. Most Mormons belong to the tribes of Ephraim or Manasseh in reference to the sons of Joseph whom Jacob predicted would grow into "a fruitful bough ... whose branches run over the wall" (Gen. 49:22). The vast majority of non-Lamanite Mormons are told they belong to the tribe of Ephraim. Native American and Polynesian Mormons, however, are more often identified as members of the tribe of Manasseh, through which Lehi traced his ancestry (Alma 10:3).

The current generation of LDS prophets holds firmly to the beliefs of their predecessors regarding Native American genealogy. During a bout of temple dedications through Latin America in 1999-2000, the church's First Presidency reminded the Saints as far afield as Ecuador, Bolivia, and Mexico of their genealogical bonds with Lehi. The potency of these reminders was amplified by their inclusion within the temple dedicatory prayers which are offered ritually before a temple becomes operational. Here is a sampling of the content of the sermons and dedicatory prayers at some recent temple dedications.

MEXICO

Colonia Juaréz Chihuahua Temple. "May the sons and daughters of Father Lehi grow in strength and in fulfillment of the ancient promises made concerning them" (President Gordon B. Hinckley, qtd. in the *LDS Church News,* 13 Mar. 1999).

Tuxtla Gutierrez Mexico Temple. "We invoke Thy blessings upon this nation of Mexico where so many of the sons and daughters of Father Lehi dwell. Bless these Thy children" (President James E. Faust, second counselor in the First Presidency, qtd. in the *LDS Church News,* 18 Mar. 2000).

Villahermosa Mexico Temple. President Thomas S. Monson, first counselor in the First Presidency, commented on the beauty of the Tabasco youngsters and reminded them that they are "children of Lehi" (qtd. in the *LDS Church News*, 27 May 2000).

ECUADOR

Guayaquil Ecuador Temple. President Hinckley noted that "it has been a very interesting thing to see the descendants of Father Lehi in the congregations that have gathered in the temple. So very many of these people have the blood of Lehi in their veins and it is just an intriguing thing to see their tremendous response and their tremendous interest." President Faust added that the "Latin people have a special quality of softness and graciousness and kindness. They are a great people—they are sons and daughters of Father Lehi, and they have believing blood. They are a beautiful people, inside and out." (qtd. in the *LDS Church News*, 7 Aug 1999)

BOLIVIA

Cochabamba Bolivia Temple. "We remember before Thee the sons and daughters of Father Lehi. Wilt Thou keep Thine ancient promises in their behalf. Lift from their shoulders the burdens of poverty and cause the shackles of darkness to fall from their eyes. May they rise to the glories of the past" (President Hinckley, dedicatory prayer, qtd. in the *LDS Church News*, 13 May 2000).

President Spencer W. Kimball, president of the church from 1973 to 1985, was unquestionably the most devoted champion of the Lamanites in the modern era. As a young apostle in 1946, he was assigned responsibility for "Indians" the world over, including those in the Pacific Islands (Gibbons 1995). He passionately carried out this responsibility, regularly reminding members of the church at General Conference of their collective responsibility to bring the gospel to the Lamanites:

> You Polynesians of the Pacific are called Samoan or Maori, Tahitian or Hawaiian, according to your islands. There are probably sixty million of you on the two continents and on the Pacific Islands, all related by blood ties. The Lord calls you Lamanites, a name which has a pleasant ring, for many of the grandest people ever to live upon the

> earth were so called. In a limited sense, the name signifies the descendants of Laman and Lemuel, sons of your first American parent, Lehi; but you undoubtedly possess also the blood of the other sons, Sam, Nephi and Jacob. And you likely have some Jewish blood from Mulek, son of Zedekiah, King of Judah. The name Lamanite distinguishes you from other peoples. It is not a name of derision or embarrassment, but one of which to be very proud. (Kimball 1982, 596)

The Book of Mormon reports that Lamanites who were converted to the Lord and became reunited with the Nephites experienced a remarkable transformation in skin color (3 Ne. 2:14-16). Their dark-skin curse was lifted and they "became white like unto the Nephites." It was once widely believed in the church that modern-day Lamanites who embraced the Mormon gospel would become "white and delightsome" just like their ancestors in Book of Mormon times. During the church's October 1960 General Conference, Elder Kimball, then a member of the Quorum of Twelve Apostles, was pleased to report this latter-day phenomenon:

> The day of the Lamanites is nigh. For years they have been growing delightsome, and they are now becoming white and delightsome, as they were promised. In this picture of the twenty Lamanite missionaries, fifteen of the twenty were as white as Anglos; five were darker but equally delightsome. The children in the home placement program in Utah are often lighter than their brothers and sisters in the hogans on the reservation.
>
> At one meeting a father and mother and their sixteen-year-old daughter were present, the little member girl—sixteen—sitting between the darker father and mother, and it was evident she was several shades lighter than her parents—on the same reservation, in the same hogan, subject to the same sun and wind and weather. There was the doctor in a Utah city who for two years had had an Indian boy in his home who stated that he was some shades lighter than the younger brother just coming into the program from the reservation. These young members of the Church are changing to whiteness and to delightsomeness. One white elder jokingly said that he and his

> companion were donating blood regularly to the hospital in the hope that the process might be accelerated. (Kimball 1960)

As scholarship relating to the colonization of the New World and Native American civilizations rapidly accumulated during the twentieth century, church leaders were anxious that LDS beliefs about the Book of Mormon not appear incompatible with prevailing views. Mormons were particularly satisfied to hear about New World scholarship that could be interpreted to support a recent Old World origin for Native American peoples. The Brethren were anxious not to disappoint, and reports of encouraging research frequently surfaced in popular LDS publications such as the *Improvement Era,* the *New Era,* and the *Ensign.* Occasionally these reports crept into General Conference addresses. In the 1960s Milton R. Hunter, a General Authority from the First Council of the Seventy, made extensive use of "Indian" accounts from the early Spanish colonial period in support of major biblical events. These included the writings of Ixtlilxochitl, a "Lamanite Mexican Prince," which seemed to corroborate the Book of Mormon so thoroughly that it was claimed to be a "Lamanite" version of the Book of Mormon. According to Hunter, Ixtlilxochitl revealed that the "first settlers to come to America following the flood came from a very high tower." Hunter cited other accounts that "verify the Book of Mormon, which claims that the ancient Americans came from the other side of the sea." As recently as 1995, Ted Brewerton, a General Authority who had served as a mission president in Central America, revived such thinking in support of an ancient origin for the Book of Mormon:

> Many migratory groups came to the Americas, but none was as important as the three mentioned in the Book of Mormon. The blood of these people flows in the veins of the Blackfoot and the Blood Indians of Alberta, Canada; in the Navajo and the Apache of the American Southwest; the Inca of western South America; the Aztec of Mexico; the Maya of Guatemala; and in other native American groups in the Western Hemisphere and the Pacific islands. …
>
> Ancient American literature contains references to a white,

> bearded god who descended out of the heavens. He is called by many names; one example is Quetzalcoatl. Historians of the sixteenth century, whose texts I have, recorded pre-Hispanic beliefs concerning the white, bearded god who came to the Americas long before the arrival of the Spanish conquerors. The following paragraphs contain examples of these beliefs.
>
> Bernardo de Sahagun (born 1499) wrote: "Quetzalcoatl was esteemed and considered as a god, and was worshipped in older times. He had long hair and was bearded. The people worshipped only the Lord."
>
> Diego Duran (born 1537) wrote: "A great man—a person venerable and religious—bearded, tall, long hair, dignified deportment, heroic acts, miracles—I affirm he could have been one of the blessed apostles."
>
> Bartolomé de las Casas (born 1474) wrote that Quetzalcoatl, the plumed serpent, was white, had a rounded beard, was tall, and came from the sea in the east, from whence he will return. …
>
> The Tamanacos Indian tribes in Venezuela have the same legend of a white, bearded god …
>
> The Book of Mormon gives an accurate account of the coming of the Lord to ancient America. (Brewerton 1995)

For many Mormons, this is as deep as their awareness of the origins of Native Americans extends. They remain oblivious to the large volume of research that has revealed continuous, widespread human occupation of the Americas for the last 14,000 years. Such research conflicts with popular LDS views patterned on the Book of Mormon and is therefore largely ignored. It is too confronting a proposition to think of Native Americans ranging over two continents 8,000 years before Adam and Eve walked the earth and 10,000 years before the arrival of the Jaredites. It is equally difficult to comprehend that indigenous Americans seem to have been unperturbed by a universal flood. These issues are acutely challenging for Mormons who accept quite literally the biblical chronology. In case members are led astray by Gentile research, the church employs apologists to vigorously defend the Book of Mormon and popular mythology surrounding it. The considerable research effort on the part of these apologists is discussed in chapters 10 through 13.

In contrast to divine assurances, popular mythology, and apologetic scholarship, the somewhat guarded and quietly acknowledged "official" position of the Mormon church is that no one knows exactly where the events narrated in the Book of Mormon occurred, only that it was somewhere in the Americas (Clark 1992). Members are not encouraged to make serious attempts to locate the lands where the Book of Mormon events occurred. This is indeed a curious position, given that Saints in all corners of the Americas are assured that they are the descendants of Lamanites. Apart from a minority group of LDS scholars, most Mormons are oblivious to the breadth and depth of Gentile insight relevant to the colonization of the western hemisphere. The disconnect between popular and scholarly views may be due in part to the fact that Mormons frequently rely on non-intellectual sources of truth when it comes to learning about the Book of Mormon.

SURE KNOWLEDGE

Among the first lessons learned by Oliver Cowdery, one of the scribes who captured the words of the ancient record as they fell from Joseph Smith's lips was how to perceive truth. Cowdery was curious about how the power to translate the gold plates into English operated, so he expressed a desire to help translate the plates, which until that time had been kept securely hidden from his view. In a revelation to Joseph Smith came the specific instructions that have since become the lifeblood of the Mormon faith:

> But, behold, I say unto you, that you must study it out in your mind; then you must ask me if it be right, and if it is right I will cause that your bosom shall burn within you; therefore, you shall feel that it is right.
>
> But if it be not right you shall have no such feelings, but you shall have a stupor of thought that shall cause you to forget the thing which is wrong ... (D&C 9:8-9)

When it comes to learning gospel truths, Mormons think with their hearts. This scripture conveys the idea that the final word on truth comes via feelings. This is now a fundamental tenet of Mormonism. It is a principal concept taught to young children entering Sunday School for the first time, and it is the principal concept in the first lesson Mormon missionaries teach those investigating the church. People are taught that if they feel good about what the missionaries teach, that feeling comes from the Holy Ghost, the third member of the Godhead, and constitutes a revelation of the truth to them. The most important object upon which Mormons apply this truth-learning formula is the Book of Mormon, just as the ancient prophet Moroni directed 1,600 years ago:

> Behold, I would exhort you that when ye shall read these things, if it be wisdom in God that ye should read them, that ye would remember how merciful the Lord hath been unto the children of men, from the creation of Adam even down until the time that ye shall receive these things, and ponder it in your hearts.
>
> And when ye shall receive these things, I would exhort you that ye would ask God, the Eternal Father, in the name of Christ, if these things are not true; and if ye shall ask with a sincere heart, with real intent, having faith in Christ, he will manifest the truth of it unto you, by the power of the Holy Ghost.
>
> And by the power of the Holy Ghost ye may know the truth of all things. (Moro. 10:3-5)

It is common to hear Latter-day Saints declare that they know with a perfect knowledge, "beyond a shadow of a doubt," that the Book of Mormon is true. Countless Mormons repeat these and similar maxims every month in testimony meetings in a tradition that begins at childhood. Once this knowledge is obtained, it follows logically that Joseph Smith must have been a true prophet and that the church that he established is literally the restored church of Jesus Christ on earth.

Since feelings are considered to be the most reliable indicator of the book's truthfulness, the same approach is considered efficacious in interpreting and extrapolating the book's message. Consequently, popular Mormon thought about who the remnants of America's seafaring Israelites are proliferates and becomes deeply implanted in the modern culture. The majority of the three million (Adherents.com 2003) Mormons in Latin America hold fast to their Lamanite heritage. If church leaders ceased reinforcing the belief today, it would continue to be sustained for years because of the momentum already in play. The message is swept along through generations attending weekly Sunday School, Institute of Religion, and LDS Seminary classes and by proselyting missionaries preaching throughout the world. Church leaders seem reluctant or powerless to curtail this belief which is so intricately woven into the fabric of the faith.

4

The Lamanites of Polynesia

We express gratitude that to these fertile Islands Thou didst guide descendants of Father Lehi.

—New Zealand Temple dedicatory prayer, 1958

Among the most precarious beliefs emerging from the Book of Mormon is the widespread assumption by Mormons that Polynesians are a branch of the House of Israel that sprouted from the descendants of Lehi. The Polynesians are believed to be the descendants of maritime Lehites who sailed into the Pacific from the Americas. Mormon folklore links the Book of Mormon sailor Hagoth, who is alleged to have lived in about 54 BC, with the colonization of the Pacific. Hagoth is said to have been an "exceedingly curious" Nephite whose talents lay in building very large ships. These were launched into the West Sea to transport large numbers of Nephite men, women, and children and their provisions to lands northward (Alma 63:5). Some of these people were never heard from again, prompting readers to conclude that the islands of the Pacific were most likely peopled when the ships lost contact with the west coast of Central or South America.

FOLKLORIC ORIGINS

Unlike most Mormon theology, which travels from the prophet down, this remarkable Polynesian belief appears to have traveled in a reverse direction, emerging first among missionaries serving in the Pacific. Joseph Smith made no recorded statements related to this belief, and in Mormon scriptures we find no explicit statement in its favor. Curiously, the predominant belief that Polynesians are Lamanites does not even square with the short Hagoth story. The Book of Mormon leaves little doubt that the adventurer Hagoth and the people who entered his ships were in fact white-skinned Nephites. Perhaps the Lamanite label stuck because the skin pigmentation of Polynesians resembles that of Laman's mainland offspring.

The first written evidence of Mormons connecting the descendants of Lehi and Pacific Islanders appears in the diary of Louisa Pratt after she joined her husband, Addison, on his second mission to Polynesia in 1850 (Douglas 1974). While preaching to a group of female Polynesians on Tubuai Island, she informed them that "the Nephites were the ancient fathers of the Tahitians," a teaching with which "they appeared greatly interested." That same year, missionaries arrived in Hawaii under instructions to teach only European immigrants. Their lack of success with the British and Portuguese prompted them to turn their attention to native Hawaiians, who were much more receptive. Within a year, missionaries in Hawaii were assuring local natives that they were members of "the House of Israel" and "God's chosen people." Among these missionaries was George Q. Cannon, who testified that ancient Israelites had sailed to America and that in time, some of their descendants had sailed west and become the progenitors of the Polynesian people (Kenney 1997). Not long after his mission to Hawaii, Cannon was ordained an apostle (1860) and then served in the church's First Presidency from 1880 to 1901.

Addison Pratt entered Polynesia in 1843 as the first Mormon missionary in the Pacific. He was a passionate sailor and had worked on whaling ships before he heeded the call to be a fisher of men. A legend that persists among some church members is that, while laboring alongside each other during the construction of the Mormon temple in Nauvoo, Illinois, Joseph Smith revealed to Pratt that the Polynesians were descendants of Father Lehi (Douglas 1974). This seems unlikely because Pratt never made the connection in his extensive diaries (Pratt 1990), nor did his missionary companions, Noah Rogers and Benjamin Grouard. It wasn't until 1858 that prophetic approval was bestowed on the speculation when Brigham Young declared, "Those islanders, and the natives of this country are of the House of Israel, of the seed of Abraham" (Douglas 1974).

Unlike the Polynesians, the inhabitants of the western isles of the Pacific did not share a strong physical resemblance with the mainland descendants of Lehi. Instead, they bore an unsettling resemblance to members of the African race, who according to revelation had inherited the harsh racial curse from God. Consequently, proselytizing among peoples in Melanesia and Micronesia became a low priority to the church. Eventually, the curse was redefined as applicable only to people who were of African descent; but in the early history of the church, missionaries were kept away from Fiji, New Guinea, New Caledonia, and the Caroline and Solomon Islands because the people had clearly mixed with darker races. From the 1868 *Juvenile Instructor*, we learn that eastern Polynesians were thought to possess the fair complexion of the "original [Lehite] stock," while "Figi [sic] ... Islanders, the New Zealanders, the inhabitants of New Caledonia and the New Hebrides appear[ed] to have greatly mixed with the Australian race or with the Negroes of New Guinea and the Philipine [sic] Islands." Consequently, Melanesia and Micronesia had virtually no contact with Mormon missionaries for the next century (Figure 4.1).

CONVERTING THE MAORI

When Mormon missionaries first reached New Zealand in 1854, most Maoris had converted to orthodox Christian sects (Barker 1967). Therefore, proselytizing focused on whites, or *pakehas*, who neverthe-

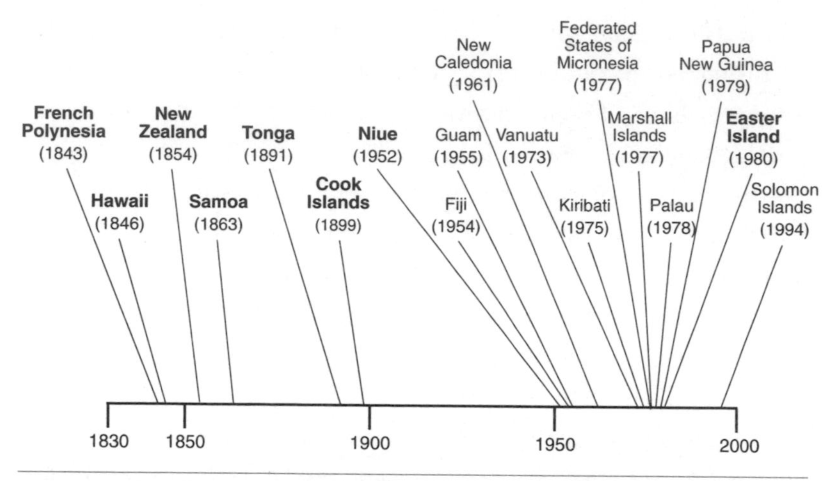

Figure 4.1 The influence of racial theology is seen in LDS proselyting in the Pacific. Missionaries first reached Polynesia in 1843 and had visited all the major Polynesian groups by the end of the nineteenth century. While missionaries first entered New Zealand in 1854, it was not until 1881 that they were specifically called to preach to the Maoris. Serious efforts to proselyte in Melanesia and Micronesia began in the 1950s, with a rapid expansion occurring toward the end of the 1970s. The church officially started baptizing Fijians and other Melanesians after missionaries consulted with anthropologists at the Fiji museum on Suva Island; Melanesians were judged to be racially distinct from Negroes (Gordon 1988). There are two Polynesian groups that were not visited until the twentieth century, but they have small populations of approximately 2,000 each. Polynesian countries are shown in bold. Adapted from the *Ensign*, Jan. 1998.

less proved resistant to the message. From 1850 until 1870, a series of wars broke out between the Maoris and *pakehas* over land, and LDS missionaries were withdrawn from the island. During this period, Maori faith in any form of evangelizing Christianity flickered, yet their faith in the Bible remained steadfast. The Maori identified with the Israelites, who had been oppressed at the hands of Gentiles. In fact with the backing of their prophets, the Maori claimed to have literal Israelite ancestry and to have migrated from Canaan at the time of the last dispersal (Underwood 2000).

In 1881 LDS missionaries were again directed by church leaders to take the gospel to New Zealand, this time specifically to the Maoris. The previous, awkward assumptions about connections to Africa were now forgotten, and within a few years, thousands were baptized. By the turn of the century, nearly one in ten Maoris were included on the church rolls. In selected settlements along the northern coast of the North Island, the percentages were higher (Underwood 2000). As with most Polynesians, the Maoris were attracted to the claim that they were descendants of Jews, which missionaries assured them of, saying that the blood of Israel ran through the Polynesians, that their ancestors had descended from Abraham and had sailed to the Pacific via the Americas. Maoris were told that, as Lamanites, they had originally possessed the gospel. This belief appealed to the native people of New Zealand because it suggested that they were members of a favored race that would one day be restored to greatness. As with all Lamanites, Mormonism "promised faithful converts prosperity in this world and with the imminent approach of the Millennium, they would be freed from the guilt of past transgressions and would receive once again white skins, and racial equality" (Barker 1967).

It is probable that the idea of a Lehite ancestry arose among missionaries because of their success with Polynesians compared to their disappointments with whites. It may have also reflected a measure of compensation for their failure to make much of an impression on Native Americans, a fact noticed by the Maori saints who in 1911 asked church leaders why Polynesians seemed to be more blessed and favored of the Lord than their American cousins. The First Presidency (Joseph F. Smith, Anthon H. Lund, and John Henry Smith) replied, saying:

> The Lord ... directed their course away from this continent [America] to their island homes, that they might not be left to be preyed upon and destroyed by the more wicked part of the House of Israel whose descendants still roam upon this continent in a fallen and degraded state. (First Presidency, 1911)

SUSTAINING THE MYTH

General Conference has been a vehicle for sustaining belief in the idea that indigenous people in the Pacific and the Americas hold a Lamanite birthright. The General Conference sermons carry a great deal of weight in the church. Words spoken there are considered to be akin to a semi-annual edition of scripture. Mark E. Petersen of the Quorum of the Twelve Apostles was particularly vocal in his support of the Polynesian Lamanite belief:

> As Latter-day Saints we have always believed that the Polynesians are descendants of Lehi and blood relatives of the American Indians, despite the contrary theories of other men. ... Recent research on the part of world-recognized scientists and scholars has focused a new light upon [Polynesians] and writings of early explorers in both America and Polynesia have become available now for detailed study.
>
> The new knowledge which has been developed shows that the Polynesians without any reasonable doubt did come from America, that they are closely related to the American Indian in many respects, and that even their traditions and genealogies bear that out.
>
> So pronounced is this feeling among the world scholars of today that one of them, Thor Heyerdahl, a widely known Norwegian anthropologist, who sailed the raft *Kon Tiki* from America to the Polynesian Islands, titled one of his books *American Indians in the Pacific*. It is a remarkable volume of great interest to Latter-day Saints.
>
> With him are other writers who confirm and re-confirm the facts now being disclosed that there is every reason to believe that the Polynesians are directly related to the American Indians, that they came from American shores and sailed westward to their Pacific Islands, and that they took with them their customs, their food, and their religion, all of which have left a permanent mark upon Polynesia.
>
> Pronounced as are these views establishing the relationship of Polynesians and American Indians, there are equally impressive data now available to disprove the theory that the Polynesians originated in the Orient and came eastward from Indonesia, Malaya, and nearby lands. (Petersen 1962)

The rest of Petersen's talk contained fragments of evidence that he felt supported the widely accepted Mormon view. The nautical exploits of Thor Heyerdahl had captured the public imagination, among whom Mormons proved to be his most enthusiastic disciples. At the time of Heyerdahl's voyages into the Pacific, they were widely publicized in the church, particularly among Polynesian members. The church-owned Liahona High School in Tonga has an extensive collection of Heyerdahl's writings and other comparable works, although few Polynesian Mormons have read his books. Still, his work is often cited in Polynesian high schools, at Brigham Young University, to missionaries during their initial training, and at church conferences (Gordon 1988). Research confirming that ancestors of the Polynesians sailed out of Southeast Asia rarely catches the attention of Mormons, and scholars who champion this view remain largely unknown among church members.

There is a familiar ring to Heyerdahl's theory of Pacific colonization. He was convinced that a race of bearded white men that originally ruled the Incan Empire built the enormous stone structures on Lake Titicaca in Peru. Over time, this wise and peaceable race was obliterated except for the high priest Kon-Tiki and his closest companions who sailed out into the Pacific. Heyerdahl believed that Kon-Tiki and his people were the founder race of all of the eastern Pacific islands. He thought that the blood of the white race was mingled with that of the North American Indians who sailed first to Hawaii and from there to all the islands to the south (Heyerdahl 1950). The flat rejection of his theories by scholars spurred him on to undertake his famous three-month journey from the coast of Peru to the Tuamotu Archipelago aboard the balsa raft *Kon-Tiki*.

Heyerdahl's research helped advance the study of Pacific prehistory, although the reasons were different from those highlighted in Petersen's sermon. Heyerdahl stimulated considerable activity among Pacific scholars to more carefully and thoroughly analyze the available archaeological evidence. This research has largely reconfirmed the strong links between Polynesia and the Orient. The iconic status Heyerdahl holds among Mormons resulted largely from selective examination of evidence in support of LDS beliefs.

Mormon leaders continue to propagate the belief that Polynesians are descended from Book of Mormon peoples. Official statements to this effect go back as far as the dedicatory prayer for the Hawaiian temple on 27 November 1919 when President Heber J. Grant thanked God "that thousands and tens of thousands of the descendants of Lehi, in this favored land, have come to a knowledge of the Gospel." President Spencer W. Kimball made repeated explicit statements that it had been his "privilege to carry the Gospel to the Lamanites from the Pacific Ocean to the Atlantic ... and in the islands from Hawaii to New Zealand" (Kimball 1975). He went so far as to propose the route the Lamanites took to Polynesia, first landing in Hawaii and then traveling south to colonize the islands of the South Seas (Clement 1980). LDS missionaries identified the Marquesas as the point of entry of the Lamanites into the Pacific (Underwood 2000) because the island was peopled in the first century, coinciding with the Hagoth migrations mentioned in the Book of Mormon. The Lamanites were assumed to have brought with them the *kumara* (sweet potato), a highly prized staple they cultivated during their sojourn in the Americas.

The Mormon belief in the Lamanite birthright of Polynesians and Native Americans has profoundly shaped the relationship between the LDS church and native peoples throughout the Pacific and the New World. On the basis of a belief in Book of Mormon ancestry, the church has invested vast sums of money in scores of church schools scattered throughout Polynesia, Mexico, and Central and South America. Other poor countries in Africa, Asia, and the neighbors of the Polynesians in Micronesia and Melanesia have been denied such liberal church investment. The church learned that community projects and educational facilities in poorer Lamanite lands could be potent factors in conversion, bringing about fulfillment of the scriptural prophecy concerning the redemption of the Lamanites (Gordon 1988). The plantation community of Laie established in 1872 on the Hawaiian island of Oahu exemplifies the success of this approach. Laie is now home to an LDS temple, a campus of Brigham Young University, and the Polynesian Cultural Center—the most popular tourist attraction in Hawaii.

Since the arrival of Mormons in the Pacific in 1843, the LDS church has been a major force in the religious life of Polynesians. There are about 300,000 Mormons in Polynesia. In no other place on earth, outside of Utah, has Mormonism penetrated so successfully. Of the nations with the highest proportion of Latter-day Saints, six of the top ten are located in Polynesia (Adherents.com 2003), including the top four: Tonga (32%), Samoa (25%), American Samoa (25%) and Niue (15%). Unlike the approach to Native American cultures, Mormonism has not sought to supplant Polynesian culture and traditions. Areas of similarity in culture and mythology have served to establish a unique bond between Polynesians and Mormons (Barker 1967), which has contributed in no small part to the blossoming of the church in that area.

The Human Race around the World

5

Human Molecular Genealogies

Behold, I will send you Elijah the prophet before the coming of the great and dreadful day of the Lord: And he shall turn the heart of the fathers to the children, and the heart of the children to their fathers, lest I come and smite the earth with a curse.

—Malachi 4:5-6

Carved straight into the mountainside of Little Cottonwood Canyon, twenty-five miles from downtown Salt Lake City, are the Granite Mountain Record Vaults, a genealogical warren containing millions of copies of microfilm, microfiche, and other records connected with the history of the human family. Constructed by the LDS church to survive whatever calamities the earth may face in the latter days, it is a repository for the original microfilm copies of genealogical records from countries all over the world. Like the Old Testament Israelites with their scrolls brimming with begats, Mormons have more than a casual interest in their ancestors.

The church invests considerable resources in helping members trace their family histories. The bulk of the genealogical work undertaken by Mormons is carried out in the central Family History Library in Salt Lake City and its 3,400 branches worldwide, housed within local meetinghouses. Much of Mormon doctrine and the perpetual work performed in temples is concerned with forging eternal family bonds by the binding ("sealing") of faithful believers to their immediate relatives and deceased forebears. While the dead are in the "spirit world" —a spiritual waiting room—living Mormons temporarily assume their names during temple ceremonies, including baptisms and eternal marriages, in order to perform these saving ordinances on their behalf. Temple work for an ancestor can only be carried out after diligent genealogical research has identified sufficient evidence of their birth, marriage, and in some cases their death.

Genealogical and temple work are concerned with welding links between fathers and mothers and their children and all of their ancestors in fulfillment of the last prophecy of the Old Testament quoted above. However, Mormons do not stop at their own ancestors. They are steadily chipping away at the family tree of the whole human race in an attempt to link all the families of the earth back via Noah to Adam and Eve, the presumed parents of the entire human family. For most Mormons, this is as far as they can comfortably venture with the human family.

Mormons demonstrate a passionate interest in people who lived in the historical period, after handwriting was developed, for whom written evidence exists of their lives. They are less concerned with the billions of people who were unfortunate enough to have died without leaving documentary evidence of their sojourn on earth. These folks must await the Millennium, a peaceful thousand-year era following Christ's return, in order to have their temple work done for them. During the Millennial dispensation, Mormons will be able to finish this work, they believe, even though it is a tall order considering the staggering number of people who have lived on the earth and the accelerating rate at which the church is falling behind. The time-consuming and detective-like obsession that genealogical research demands does not fit well with the heavy church responsibilities of many Mormon families. Consequently, most of the people sitting at microfilm readers in Mormon Family History Libraries are not Mormons. The extensive resources, polite and helpful staff, and the minimal cost of using the libraries attract many Gentiles who, in fulfillment of Malachi's prophesy (Mal. 4:6), have turned their hearts to their fathers.

Fascination with understanding the human family tree drives research within scientific disciplines, as well, and the curiosity extends far beyond tracing our immediate ancestors. Over a century of research by archaeologists and anthropologists has yielded a large body of information about how humans have evolved, how human civilizations and cultures developed, and how people have migrated around the world.

OUR FAMILY HISTORY

For scientists, the human family tree is actually a branchlet on the tree of life. The genus *Homo* is a small branch on the primate and then mammal branch that joins the common family tree of all living things. The potent truth that plant and animal species have evolved from earlier, simpler forms of life lies at the heart of Charles Darwin's *On the Origin of Species,* published in 1859. While the principle of evolution by natural selection has been hotly debated among scientists, religionists, and philosophers, it remains the central unifying principal of modern biology, backed by evidence from numerous fields in the natural sciences. The most vehement critics of evolution have insisted that humans arrived on the earth via the independent process of a special creative act, contrary to the evidence accumulated to date indicating that humans were part of the evolutionary process. But even Darwin noted that the evolutionary viewpoint need not imply the absence of a creator:

> There is grandeur in this view of life, with its several powers, having been originally breathed by the Creator into a few forms or into one;

> ... from so simple a beginning endless forms most beautiful and most wonderful have been, and are being evolved. (Darwin 1859)

Early attempts to understand the hereditary relationships among different human populations were usually dashed on the rocks of orthodox Judeo-Christian dogma. Even the most far-flung members of the human family had to be accounted for within the strict biblical time frame. People of all races and locations on the earth were said to be descended from the family of Noah, who stepped from the ark 4,500 years ago. In contrast, scientific research accumulated evidence that significant human populations lived uninterrupted over much of the earth for tens of thousands of years. Among the most notable discoveries in the field of human genealogy has been the genetic evidence linking related living populations to ancestral populations who undertook ancient migrations. This has been particularly true for people living in the quarter of the earth most recently discovered by Europeans, the lands in or adjacent to the Pacific Ocean.

Most scientists agree that all human populations on the earth originated from a relatively small founding population that dwelt in Africa between about 50,000 and 200,000 years ago (Foley 1998). About 50,000 years ago, our ancestors began making rapid cultural advances, probably brought about by the development of communication skills (Diamond 1997). One of the earliest dispersals out of Africa was apparently around the northern rim of the Indian Ocean into Southeast Asia and Papua New Guinea, reaching Australia by about 50,000 years ago. From about 45,000 years ago, a number of dispersals led to Europe and into eastern Asia. Areas of European and Asian colonization expanded and contracted over the next 30,000 years, largely in response to advances and retreats of glacial ice. During this period the closely related archaic *Homo* species became extinct, including the Neanderthals in Europe and probably *Homo erectus* in Southeast Asia. Approximately 20,000 years ago, the world's climate dipped to its coldest in the last 130,000 years. Lowered sea levels at the peak of the last Ice Age exposed a number of land bridges around the globe. These allowed human populations to walk between Asia and America and between Papua New Guinea, the Australian mainland, and Tasmania.

About 10,000 years ago, agriculture gradually began emerging in various places around the world. Until this time our ancestors had lived as small groups of related individuals supporting themselves by hunting and gathering. Domestication of plants and animals underpinned dramatic population growth and the establishment of the first permanent settlements. Wild relatives of many of the first domesticated cereals such as maize, wheat, and barley have been found in proximity to some of the earliest human settlements (Heun et al. 1997). In fact, the age of some of the earliest human settlements has been ascertained by radiocarbon dating crop seeds littered about ancient ruins.

The fossil record has revealed much about our origins, including where our ancestors lived and their relationships to living descendants. However, fossils are hard to find and frequently of poor quality, making it difficult to obtain accurate measurements of relationships. The environment can also affect the characteristics of skeletal remains. Increasingly, scientists have turned to genetic information as a means of comparing human populations. Genetic evidence is abundant and easy to collect, and vast amounts have been accumulated for human populations the world over. More recently, information from the field of human genetics has allowed researchers to create molecular genealogies. The genealogy written into our genes provides scientists with molecular pedigrees to, not only trace our relationships with each other, but our relationships with all living things beyond Adam.

In the middle of the last century, scientists began studying the protein products of our genes, particularly blood proteins, in order to infer relationships among human populations. The most well-known blood protein used for this purpose is the one that determines blood groups, an important genetic trait to know before a blood transfusion. Human populations can be characterized on the basis of the frequency of the different kinds of blood-type proteins (alleles or gene variants)

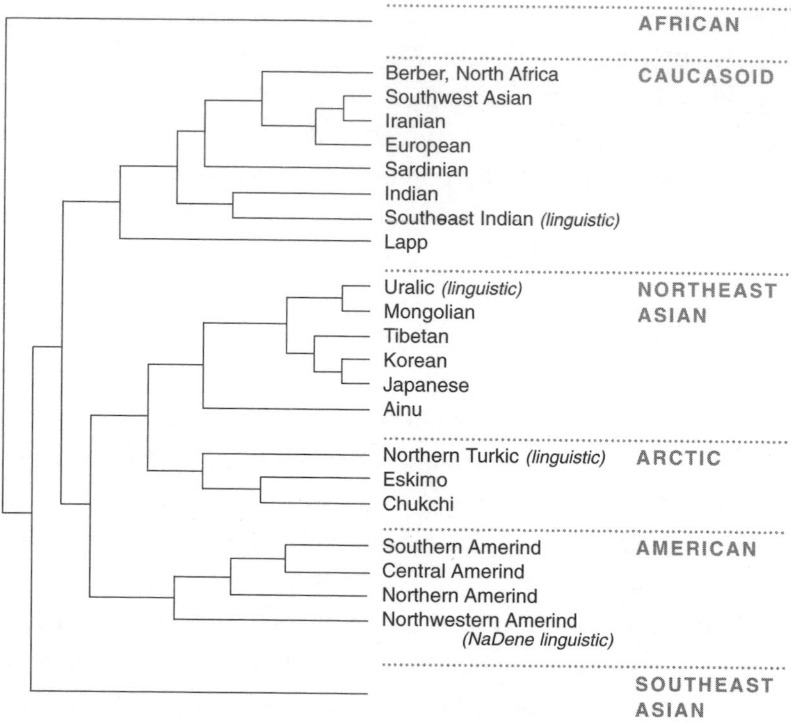

Figure 5.1 Human populations are distinguished by thé frequencies of particular protein variants carried by individuals who make up those populations. This protein variation reflects genetic variation in each population. In this figure, the shortest lines join closely related populations with similar patterns of protein variation. Much longer distances and deeper branches separate distantly related populations. The position of the different populations in the family tree reflects patterns of human migration around the world. Redrawn from Cavalli-Sforza et al., 1988.

A, B, and O among its individuals. For example, the O protein variant is found at high frequencies among Native Americans where it approaches 100 percent (Cavalli-Sforza et al. 1994). If the variants in large numbers of different proteins are collectively examined in numerous global populations, it allows identification of related populations which naturally share similar patterns of protein variation. A vast array of this type of genetic information, used to reveal relatedness between human populations, has been assembled in *The History and Ge-*

ography of Human Genes, edited by L. Luca Cavalli-Sforza and colleagues in 1994.

HUMAN DNA

When DNA was discovered about fifty years ago, geneticists reached the greatest source of information that can be used to characterize populations. DNA encodes our genes, which direct the synthesis of proteins that carry out the work of making us who we are. The variation in proteins that scientists previously examined results from differences in the sequence of DNA letters that code for those proteins. DNA contains abundant information which can be used for measuring the genetic likeness of people in different areas of the world.

DNA is a long chain-like polymer comprised of four repeating units called bases, represented by the letters A, C, G, and T. Hereditary information is stored in coded form in the order of these lettered bases. There are roughly 3 billion bases in the human genome, and every cell in our body contains an identical copy. DNA superficially appears to be a random arrangement of the four-letter genetic alphabet; however, within long stretches of sequence, we find short segments with specific functions—genes. We share the same basic library of genes. In fact, everyone has a duplicate set—one derived from each parent. The molecular library with the complete set of instructions for constructing the human body comes in forty-six volumes called chromosomes—twenty-three chromosomes from our mothers and twenty-three from our fathers.

We inherit genes as shuffled arrangements of the chromosomes from earlier ancestors. The shuffling process is known as recombination and occurs every generation, resulting in complex patterns of chromosomal inheritance. A person's genetic "hand" reflects what that person has been randomly "dealt" from his or her ancestors going back many generations. Identical twins are dealt exactly the same hand, while all other offspring contain one among an infinite number of possible "hands." Hence, few siblings appear exactly alike. This genetic shuffling makes it difficult to identify which forebear a particular gene came from. Due to this limitation, many geneticists examining human genealogy now study genes that escape the confusion of recombination. The most widely studied are the genes contained in the mitochondria and Y chromosomes.

Mitochondria are the powerhouses of the cell. They constitute minute compartments that are specialized for the production of energy (Figure 5.2). Their size is similar to that of some bacteria, but their resemblance to these microscopic creatures does not stop there. At a distant time in our long history, evolution resourcefully recruited a bacterium to help run the cell. The clearest sign of this is the "DNA language" that our mitochondrial genes speak, which is clearly related to the genetic language of some of our very long-lost bacterial cousins. Over time, the number of mitochondrial genes has been trimmed down to just thirty-seven, with a few of the original genes incorporated into the chromosomes in the nucleus. An important advantage of having mitochondria is the capacity to vary energy production in different cells by varying the number of mitochondria. Heart muscle cells have enormous energy demands and contain thousands of mito-

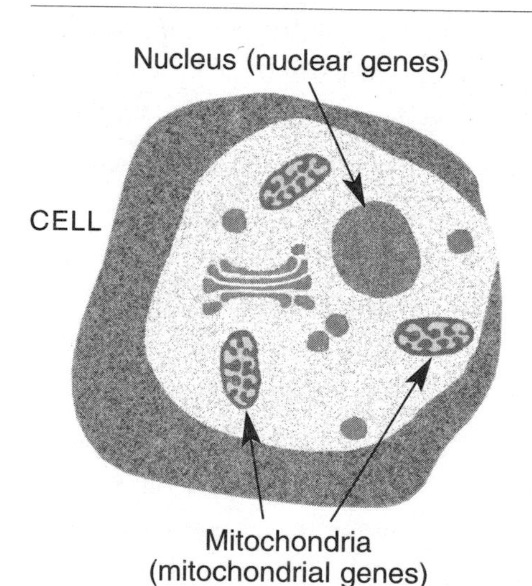

Figure 5.2 Where are our genes? Every one of our billions of cells has an identical copy of our DNA containing the blueprint for all our genes. Over 99.9 percent of our genes are found within the nucleus of our cells, which is where our chromosomes are located. A tiny fraction of our genes is contained within structures called mitochondria, specialized compartments for synthesizing energy for the cell. Each human cell contains 5,000-10,000 mitochondria. Some cells have a high energy requirement, such as muscle cells, and have an even larger number of mitochondria.

chondria. Additionally, each mitochondrion (singular) contains many copies of the mitochondrial genes. Both of these factors ensure the capacity to synthesize large amounts of the enzymes necessary for energy metabolism.

If we were to compare all the DNA letters in our genes to a 500-page novel, the strand of DNA in mitochondria, which contains about 16,500 letters, would barely reach the end of a single five-letter word. This is why it has taken so long to sequence the entire human genome when the mitochondrial DNA sequence has been known for about twenty-five years. The Y chromosome contains several thousand times the amount of DNA found in the mitochondrial genome. Many of the genes on the Y chromosome are necessary for the development of male physical characteristics, but clearly half the human population finds them unnecessary.

HUMAN MOLECULAR GENEALOGY

Mitochondrial DNA and Y chromosome DNA are useful for studying human populations because they remain remarkably intact from generation to generation. Human mitochondrial DNA is maternally inherited; the mitochondria in sperm cells do not penetrate the egg upon fertilization. Males receive their mitochondrial DNA from their mothers but cannot pass it on to future generations. Human Y chromosome DNA escapes the complex chromosomal rearrangements in each generation because it is passed from father to son as a single entity. As a consequence of these unique modes of inheritance, the mitochondrial DNA represents a single maternal line and the Y chromosome represents a single paternal line. Both are passed down through the generations like maternal and paternal surnames.

For the most part, the order of the DNA letters contained within the mitochondrial and Y chromosome DNA remain fixed as they are passed from generation to generation. However, in practice they gradually accumulate copying errors and radiation-induced errors that slightly alter the sequence of letters. The result is the development of branching maternal and paternal pedigrees that descend from earlier common lineages. Branches in the tree arise where new mutations enter a particular DNA lineage and are passed on to future generations. Closely related DNA lineages typically have very similar sequences and share the same mutations.

Scientists use the term "haplotype" to denote an individual DNA lineage and "haplogroup" to denote a group of related haplotypes. Each haplogroup is identified by a letter of the alphabet, such as "H," "B," or "X." Throughout this book, I will generally use the terms lineage or line when referring to a haplotype and the terms lineage family, group, or cluster when referring to a haplogroup. Occasionally I refer to a specific haplogroup as a lineage or line, as in the X lineage.

Studying the DNA lineages for humans reveals a lot about how mankind migrated around the world. Each migratory party typically possessed a subset of the DNA lineages of the parent population. DNA mutations that arose before humans spread throughout the world are often found in widely separated populations. Changes that occurred during later stages of colonization are usually found only in particular regions and continents. Consequently, the DNA lineages in a population represent what was bequeathed them in the way of random changes in the DNA of its ancestors, as well as changes among its own individuals. The global distribution of maternal and paternal lineage groups reflects the interplay between these two influences (Figure 5.3). For both women and men, Africa contains the greatest diversity of DNA lineages, as would be expected for the birthplace of the human race.

Mitochondrial lineage analysis has made a remarkable contribution to the study of genealogy. Scientists recently succeeded in extracting and analyzing DNA from the fossilized bones of a Neanderthal (Krings et al. 1997), a cave-dwelling relative of modern humans. Neanderthals lived in Europe and western Asia from around 230,000 years ago until about 30,000 years ago. The DNA was isolated from an arm bone originally uncovered in 1856 by quarrymen in the Neander

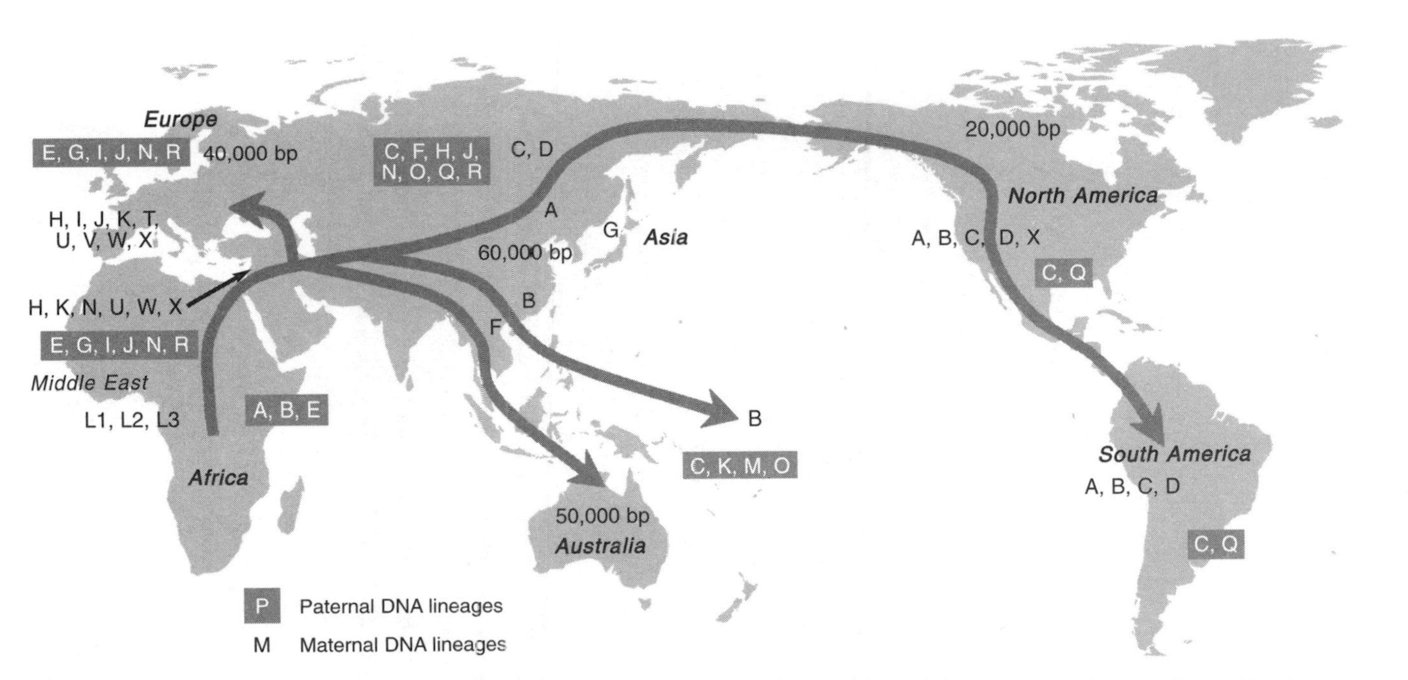

Figure 5.3 Distribution of the major maternal and paternal DNA lineage families (haplogroups) throughout the world. The paternal and maternal family trees are rooted in sub-Saharan Africa, from where they spread across the globe. The positions of lineages on the map indicate where they occur exclusively or at the highest frequencies. Unboxed letters indicate female lineages; boxed letters indicate male lineages. The dates indicate the approximate time of arrival as BP (before present) for each continent. Male lineages are as designated in Hammer and Zegura 2002, female lineages as widely accepted.

Valley near Düsseldorf in western Germany. When the sequence of about 380 bases of Neanderthal mitochondrial DNA was compared to a collection of human sequences, the differences were far greater than between any two human sequences. There were also no Neanderthal-like mitochondrial DNA sequences among the thousands of human samples sequenced to date. Other researchers have since obtained similar results with other Neanderthal specimens. The number of differences between humans and Neanderthals is about half the number occurring between humans and our closest living relatives, chimpanzees. The results provide strong evidence that the Neanderthal was not our direct ancestor. Instead, we shared a common ancestor hundreds of thousands of years earlier. It would appear that during the thousands of years that our direct ancestors shared Europe with Neanderthals, both groups remained reproductively isolated.

Glimpses into our more recent evolutionary history have come from looking at the paternal and maternal components of our heritage. Most geneticists and paleontologists believe that modern humans originated from a small population that lived in Africa between 100,000 and 200,000 years ago. Both the Y chromosome and mitochondrial DNA are useful for asking questions about models of human migration and allow scientists to track both sexes independently through their genes. The aim of this type of research has been to determine when humans undertook major global migrations, particularly the early emigration out of Africa. Two important questions are frequently considered: when did all humans share a common ancestor and when did humans begin migrating away from Africa?

A detailed look at human mitochondrial variation was published in 2000 by Max Ingman and colleagues, who obtained the entire 16,500 odd DNA letters in the mitochondria of fifty-three people who lived in diverse global locations. This analysis arrived at the conclusion that humans had a most recent common female ancestor in sub-Saharan Africa about 171,500 years ago. They estimated that African and Asian populations first split from each other about 52,000 years ago. A parallel study examined a small region of the X chromosome that escapes generational mixing. This work yielded an estimate of about 178,000 years ago for our common female ancestor (Kaessmann et al. 1999).

The molecular lineages of men are rooted in Africa as well. In one study of over 1,500 Y chromosomes from global populations, it was estimated that the most recent male ancestor of all modern humans lived in Africa approximately 150,000 years ago (Hammer et al. 1998). A more recent study suggests that men shared a common patriarch as recently as 90,000 years ago (Hammer and Zegura 2002), a date more compatible with paleoanthropological and other genetic research on human origins. Rather than suggesting an origin in a single man and woman, such as an Adam and Eve, it is more likely that we descended from a population of several thousand individuals. It is now widely accepted within numerous areas of research, including paleontology, archaeology, anthropology, and genetics, that the human family tree sprouted from such a group in sub-Saharan Africa.

Molecular genealogy can also be used to address questions arising about our more recent ancestry. As soon as Y chromosome lineage studies were feasible, Brian Sykes, Professor of Human Genetics at Oxford University, took the opportunity to delve into his own paternal family history (Sykes and Irven 2000). He isolated DNA from cheek swabs posted to him by forty-eight fellow male Sykeses living in West Yorkshire, Lancashire, and Cheshire, counties encompassing the towns with the earliest occurrences of the surname. About 44 percent of the Sykeses tested shared an identical Y chromosome not found in a group of unrelated male neighbors or a group of randomly selected men from all over England. This discovery revealed the close ties between the surname and DNA for the 44 percent with identical Y chromosomes. It also uncovered evidence of a significant level of non-paternity, estimated to be about 1.3 percent per generation. In other words, a significant proportion of Father's Day cards in the Sykes family miss the mark.

The way in which molecular genetics can trace the invisible flow of genetic information from generation to generation has captured the attention of human genealogists tracing our recent ancestors and of anthropologists probing further into our deepest past. Among the fields of research benefitting the most from this technology are those that seek to understand the origins, timing, and nature of the human migrations to the western hemisphere and the islands of the Pacific.

6

Science and the First Americans

Lo! the poor Indian, whose untutored mind sees God in clouds, or hears him in the wind.

—Alexander Pope, 1733

It would be somewhat boorish to launch into a discussion of Native American DNA genealogy—the subject of the next chapter—without surveying the scientific landscape into which modern molecular biology research has intruded. It came as no surprise to most scientists to learn that the DNA of living indigenous Americans was most homologous with the DNA of Asians. Well before the structure of DNA had been determined, the Asian source had been accepted through the steady accumulation of over a century's worth of research from many disciplines. It was, and still is, widely accepted that the first waves of colonization occurred around or before 14,000 years ago from Siberia by way of the Bering Strait. Further contacts with Asia appear to have been limited to contact between hunter-gatherers adjacent to the Bering Strait and the possible movement of people accompanying the sweet potato as it moved from South America to Polynesia. Apart from the Norse who briefly stepped into Newfoundland, there is little evidence of contact with Europe before Christopher Columbus came face to face with the native inhabitants of the Caribbean.

Widely held scientific views inevitably spawn the formulation of alternative theories, and this is particularly true regarding the colonization of the New World. Many of these theories hinge on a belief that large advances in technology underpinning major civilizations were unlikely to have occurred at the same time in different places around the world. Some have argued that the Old World was the source of technological ideas that were disseminated from there to the rest of the world—diffusionist theories that rely on the assumption of marathon ocean voyages to the New World. Such alternatives have been attractive to Mormon scholars, who have been among the fiercest critics of mainstream views.

NEW WORLD CHRONOLOGY

Two twentieth-century scientific breakthroughs deserve mention at the outset—dendrochronology and radiocarbon dating—because they revolutionized New World scholarship. Both permitted development of a clear chronological framework into which all New World archaeological research could be positioned. In 1909 the American astronomer Andrew Douglass recognized that the annual growth rings of certain trees reflect historical climatic conditions at their fixed location (Giddings 1962). They formed wider growth rings in favorable years and narrower ones in lean years. Using ring measurements taken from living and dead California bristlecone pines with overlapping ages, Douglass constructed annual climatic records dating as far back as 7,000 years before the present. His real insight was to match the rings in wood from ancient American Indian ruins to existing tree ring chronologies, thus determining the age of the timber ruins. Douglass first applied this technique, now called dendrochronology, to date roof beams of Anasazi dwellings in the Four Corners region in Arizona, New Mexico, Utah, and Colorado. The Pueblo Bonito cliff dwellings in Colorado were found to have been inhabited from AD 1000 until they were abandoned in AD 1300 (McGraw 2000). Dendrochronology has been widely used since to date structural timbers in prehistoric buildings and boats from around the world.

Radiocarbon dating is arguably the most important technology to have been applied to modern archaeology. Pioneered by Willard Libby at the University of Chicago in 1948 (Libby 1955), this method makes use of the unique chemistry of carbon to measure the age of organic, carbon-containing material. A small fraction of atmospheric carbon dioxide contains radioactive carbon-14. Both varieties of carbon enter all living things, plants during photosynthesis and animals after eating plants. After plants and animals die, about half the carbon-14 they contain decays to nitrogen every 5,800 years. Libby reasoned that we could estimate the age of ancient organic materials—wood or bones—by measuring the proportions of carbon and radiocarbon they contained. If steps are taken to avoid contamination with different-aged material, such as younger material from invasive tree roots or older material such as coal in a fireplace, dating to within 5 percent accuracy can be achieved for material between 500 and 50,000 years old. Pre-European New World archaeology fits comfortably within this window of utility. Scientists frequently talk in radiocarbon years when referring to the age of artifacts and the chronology of New World archaeology; however, I have used calendrical years throughout.

An early technical speed bump encountered by scientists was the observation that atmospheric carbon-14 levels have not been constant throughout the earth's history. This problem was circumvented by measuring carbon-14 levels in tree rings that already had been dated precisely by dendrochronology, thus allowing accurate compensation for this natural fluctuation. While radiocarbon dating can only date organic remains, the consecutive layering of the remains of past civilizations at sites of continuous settlement permits dating of non-organic materials occupying the same layer. This principle of stratigraphical observation is a cornerstone in modern archaeology that was already well understood by the gifted polymath Thomas Jefferson in 1784 when he carefully excavated Indian mounds on his estate in Virginia.

THE FIRST AMERICANS

The term New World is a particularly apt description for the Americas, but for reasons different from those that inspired Old World Europeans to coin it. For most of the time period that humans or creatures closely resembling us have existed on the earth, they have not inhabited the Americas. The earliest human remains in the Americas are barely 13,000 years old (Powell and Neves 1999). No trace has been found of the Neanderthal, *Homo erectus,* or any of our other prehistoric relatives. When humans entered the vast American landmass, it was home to a wide diversity of animals, some of which had only recently emigrated from Asia. However, the Americas were a place where mankind had never before lived.

Most scholars consider that the first immigrants were family groups of nomadic Siberian hunters who followed herds of bison and mammoth across the bleak 1,000-kilometer-wide wilderness that joined the two continents. It is unlikely that they had the slightest notion they were entering a continent where mankind had never before set foot. Asia and the continent of North America are currently separated by the Bering Strait, a treacherous stretch of water about sixty miles across at its narrowest. At the peak of the Wisconsin Ice Age, 18,000 years ago, world sea levels were more than 300 feet (91 meters) below their present levels, exposing a wide and flat land bridge known as Beringia that linked Siberia and Alaska. The drop in sea levels occurred as large amounts of the earth's water became trapped in polar ice sheets. About 10,000 years ago, as the earth was warming, the land bridge was resubmerged and the two continents became detached (Fagan 1987). Since then, the Bering Strait has hindered movement between the continents. It is not an absolute barrier because, even now in the most severe winters, ice forms between the two continents that is solid enough to walk on (Crawford 1998).

At the dawn of the twentieth century, the prevailing view was that the Americas had been inhabited for only a couple of millennia. This view was seriously compromised with the discovery in the 1920s of a primitive stone spear point embedded between the rib bones of a bison at Folsom, New Mexico. This particular species of bison was known to have been extinct for 10,000 years or more. Similar primitive stone points were found nearby several years later alongside the bones of mammoths in Clovis, New Mexico (Figure 6.1). There could be little doubt that humans had been in the New World since the retreat of the last glacial ice sheets. Pushing back the time of entry by several thousand years relieved scientists of having to account for the diversity of Native American cultures and languages within such a tight time frame. As many as 1,500 different languages were spoken by New World tribes that greeted the early European immigrants. Extending the period of occupation also accounted for the few surviving similarities between New World cultures and the Asian cultures from which they were widely thought to have emerged.

These early Americans are now known as the Clovis people, highly skilled hunters who pursued Ice Age mammoths, camels, bi-

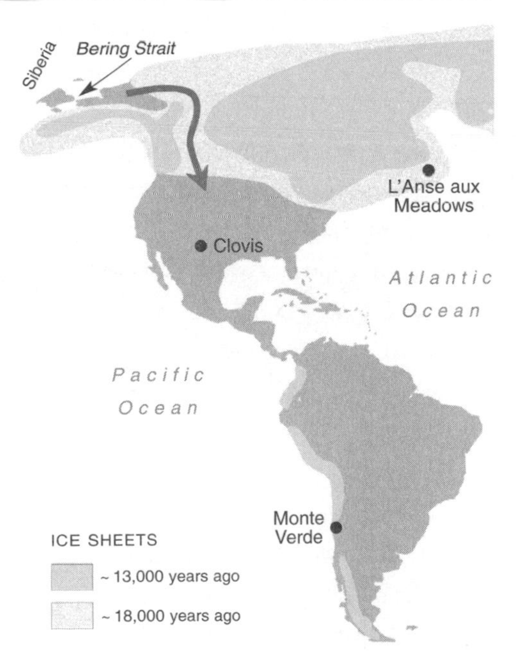

Figure 6.1 Until about 13,000 years ago, Alaska was cut off from North America by continental ice sheets. The map shows the approximate position of the ice sheets over North and South America during the Wisconsin Ice Age and the ice-free corridor that was open approximately 13,000 years ago. The locations of some of the important archaeological sites mentioned in this chapter are also shown.

son, and horses about 13,000 years ago over much of North America. Within a few millennia of the arrival of the Clovis people, many of the larger mammals they harvested began to disappear, including the mastodon elephant, saber-toothed cat, horse, several species of camel, giant sloths, and other animals (Grayson 1987). Some argue that climatic changes at the end of the Ice Age contributed to the decline of these species, but the fact that these beasts had previously survived numerous Ice Ages by simply walking to warmer climes and the exact correlation with the entrance of man suggest that they were eaten into extinction (Diamond 1997). In Australia and the Pacific, the demise of large animals similarly coincided with the arrival of man.

Just as the date of Christ's birth divides the Christian world's calendars, Clovis has become a defining point in American archaeology. The culture appears suddenly in the archaeological record and blankets the continental United States (the southern forty-eight states) and extends into Mexico. Radiocarbon dating places the Clovis culture firmly in the period 13,000-13,500 years before the present. For the last half-century, archaeologists have sought and disputed evidence of a human presence in the Americas predating the Clovis people. Scores of pre-Clovis claims have come and gone, failing to weather the criticism of an increasingly skeptical archaeological community. Among the sites currently feeling the heat of further scrutiny is Monte Verde in southern Chile (Figure 6.1). Nearly thirty radiocarbon ages from wood, charcoal, and ivory place the age of this small human settlement at about 14,500 years before the present (Dillehay 1997). Several seasoned critics have been persuaded by these discoveries, but others remain unconvinced. There is a lot at stake. It is generally believed that the migration routes from Alaska through glacier-choked Canada were virtually impassable (Figure 6.1) between about 13,000 and 20,000 years ago (Jackson and Duk-Rodkin 1996). If sustained, the findings at Monte Verde suggest an earlier arrival in North America, probably more than 20,000 years ago. Scientists continue to search for signs of human activity before the blossoming of the Clovis people, and heated debates about when people first entered the New World are unlikely to cool for some time.

The most conspicuous clue that the aborigines of America entered the New World from Asia is that people from both territories bear such a strong physical resemblance to each other. Many native Siberians would go unnoticed at a Native American powwow if suitably attired. The two groups share morphological features characteristic of Mongoloid peoples such as straight black hair, prominent cheekbones, sparse body hair, reddish to brown skin, relatively flat faces, and the Mongoloid sacral spot (Crawford 1998). Other frequently shared traits are darkly colored eyes, the Mongoloid eye fold, and dental characteristics such as shovel-shaped incisors. Not all Native Americans—or Asians, for that matter—possess all of these traits, and there is considerable variation within and between populations. Several factors probably contribute to this variation. The first Americans brought a small fraction of the variation present in the Asian source population. Differences arose in the American populations and among their relatives in northeastern Asia since separation. Considerable gene flow from Europeans and Africans has since masked some of the pre-Columbian morphological traits of some native populations.

While the Asian influence on contemporary Native Americans is widely acknowledged, the evidence is growing for the idea that the earliest immigrants were different in appearance from those who followed. Scientists have noticed that the oldest skulls recovered in the Americas do not look like contemporary Asians or American Indians; instead, they look more Caucasian. In fact, they most closely resemble the Ainu (Preston 1997), the earliest colonizers of Japan who now reside on the island of Hokkaido. Among the most famous of these skulls was discovered on the banks of the Columbia River at Kennewick, Washington, in 1996. As scholars attempt to solve the puzzle of the origins of Kennewick Man and his early companions, some have looked to Europe as a possible origin for the earliest Americans. There are cultural similarities between the Clovis in America and the Solutrean people who inhabited France and Spain approximately 16,000 years ago and no similar evidence for similarities with Asian culture (Preston 1997). At that time in history, the Ice Age was almost at its maximum, and the North Atlantic was almost certainly frozen from Norway to Newfoundland. While it is possible that migration occurred from Europe, it would have been an extremely perilous crossing. Most archaeologists favor the idea that if Caucasian peoples were among America's founders, they came via Asia.

There is widespread agreement that most ancestors of the present-day American Indians arrived via Beringia, but there is considerable debate about the timing and number of migrations that took place. Before the discoveries at Monte Verde, many scholars thought the ancestors of all Amerindian tribes crossed the Bering land bridge in essentially a single migration about 13,000 years ago. However, this date has become tentative in light of the research in South America and elsewhere (Adovasio 1983, Guidon and Delibrias 1986, Roosevelt et al. 1996). Another popular theory argues that American populations originated from three sequential migrations of northeast Asians. This migration scenario was developed on the basis of comparisons of dental morphology, linguistic associations (Turner 1983, Greenberg et al. 1986), and blood protein variation (Williams et al. 1985). According to this hypothesis, the first migrations took place 15,000-30,000 years ago and gave rise to the Amerindians (Paleo-Indians) who occupied most of North America and then rapidly colonized most of Central and South America (Griffin 1979). More recent migrations gave rise to the Na-Déné Indians of the Pacific Northwest (~10,000-15,000 years ago) and the Aleut-Eskimo (Griffin 1979, Hopkins 1979). It has been noted that there may be support for this three-wave, north-south migration in gene distribution patterns among Native Americans (Cavalli-Sforza et al. 1994).

NEW WORLD CIVILIZATIONS

The earliest settlers in the Americas were primarily small groups of hunter-gatherers who likely developed few significant permanent settlements. Their total dependency on animals and wild plants for food bound them to a nomadic existence, following the herds of animals they hunted. In parts of the New World, however, the hunting-gathering way of life was gradually complemented by the domestication of indigenous plants and animals as early as 10,000 years ago.

New World attempts to domesticate animals were frustrated by the fact that most potential domesticates had recently become extinct. There were few fast growing, herbivorous, non-panicky mammals with a kindly disposition and inclination to breed under the watchful eye of human captors (Diamond 1997). An animal that lacked any one of these traits was bound to stymie even the most dogged attempts at domestication. Only one of the fourteen large herbivorous mammals that humans have domesticated originated in the Americas. All of the "big five" (sheep, goats, horses, cattle, and pigs) were domesticated in the Old World. Only four species of animals native to the Americas made it into the livestock ranks, including the llama as a pack animal, the alpaca for its wool, and the guinea pig and turkey for food. The llama and alpaca are thought to be well-differentiated breeds that originated from the same wild species just as the variety of modern dog breeds sprang from common parents.

The first domesticated plant in the New World was the not-so-humble potato, tamed in the Lake Titicaca basin region of the Andes in South America (Smith 1994). At least four species of tuber-producing crop plants were domesticated, the potato being the most widely grown. Other Andean crops included varieties of beans, chenopodium, and cotton, while elsewhere in the Americas, native people domesticated maize, beans, and orange pumpkin squash in Mexico and sunflowers, chenopodium, and green and yellow squash in northern and eastern North America. Cassava, an important tropical crop, was domesticated along the southern border of the Amazon Basin adjacent to the border of Brazil and Bolivia (Olsen and Schaal 1999).

Maize was the American staff of life and central to the lives of many Native Americans. It is widely agreed that it was a domesticated form of teosinte (Figure 6.2), a wild Mexican grass of strikingly dissimilar appearance (Doebley et al. 1984). DNA fingerprinting of a range of teosinte grasses has revealed that the varieties most closely related to modern maize grow in the Balsas River basin in southwestern Mexico, suggesting that this is near where it was first domesticated (Wang et al. 1999). The archaeological record reveals the steady evolution of maize over the millennia as indigenous farmers "naturally selected" superior varieties. Maize cultivation gradually spread north and south from Mexico. In combination with beans, which are nutritionally complementary, maize underpinned considerable population growth in the Americas and the development of major civilizations.

Most of the principal domesticated plants were sub-tropical or tropical in origin—maize, potato, beans, chili peppers, cassava, and squash—and known to have a low frost tolerance. This was a critical factor in setting the limits of New World population expansion. The largest populations were in the frost-free regions that permitted year-round cultivation of crops. The most noticeable cultural change that

Figure 6.2 The American staff of life, maize (*zea mays*) is a highly domesticated form of a wild Mexican grass named teosinte. The latter occurs naturally in the Balsas River Valley in southern Mexico. The two plants are different enough in appearance that at one time teosinte and maize were not classified in the same genus. But genetics research suggests that mutations in as few as five genes are responsible for the dramatic changes in maize that made it so suitable for agricultural production. A mutation in a single gene has been recently shown to cause a striking change in teosinte architecture so that it closely resembles maize.

accompanied the emergence of agriculture was the establishment of the first permanent settlements. The earliest New World civilization, the Olmec, lived in the humid lowlands of the Mexican Gulf Coast. They were superseded by other cultures including the three major empires that blossomed in the New World—the Mesoamerican Aztecs and Mayans and the South American Incas—all underpinned by sophisticated agricultures.

Two Mesoamerican civilizations, the Olmec and Maya, merit a brief sketch because, for decades, they have enthralled Mormon archaeologists, whose expertise has traditionally focused on these cultures. Mormon scholars frequently correlate the Olmec and Mayan civilizations with the Book of Mormon Jaredite (2200 to 600 BC) and Nephite (600 BC to AD 400) civilizations respectively, the two major groups about which the Book of Mormon is principally concerned.

The Olmec farmed the fertile, Nile-like levees in southern Veracruz and the neighboring state of Tabasco from about 1500 BC until their decline in 400 BC. They are most famous for the colossal stone heads—multi-ton portraits of their kings—that they carved from basalt imported from great distances. The Olmec civilization relied heavily on maize, which subsequently contributed about four-fifths of the nourishment of Mesoamericans. An important development that may have nudged the Olmec civilization into existence was the development of a process for cooking maize known by the gaudy word "nixtamalization," a method invented by the Olmec to soften the grain's hard external pericarp (Coe and Coe 1996). The process, which involved cooking maize with alkali-containing substances such as white lime or wood ashes, dramatically enhanced the nutritional value of the grain. We owe a great debt to the Olmec who, in addition to bringing maize to prominence, most probably domesticated *Theobroma cacao*—the chocolate tree—in their tropical homeland.

The later "Classic" Mayan civilization was influenced by the Olmec and flourished from about AD 250 until its precipitous decline after AD 800 (Coe et al. 1996). The Mayas dotted the lowlands of the

Petén or Yucatan Peninsula with their towering temple pyramids and magnificent palaces and plazas. They developed complex cultures with elaborate calendrical and astronomical systems. It was among the Maya that the Mesoamerican hieroglyphic writing reached its highest elaboration. Most hieroglyphics were written on perishable bark-paper codices, of which only a handful survived the Maya collapse and Spanish bonfires. Surviving Mayan texts are also found on wall paintings, stone reliefs, and delicately carved and painted ceramics. Most Mayan writings have now been deciphered and reveal that the cultural giant was wracked with frequent internal warfare between competing city-states. Blood-letting, torture, and human sacrifice, particularly of enemy captives, were fundamental religious rituals in Mayan society. The dramatic collapse of this empire was most probably triggered by overpopulation and severe environmental degradation.

By the fifteenth century, large New World empires, supported by populations of approximately ten to twelve million, were found in both Mexico (Aztec) and Peru (Incan) (Cavalli-Sforza et al. 1994, Fagan 1987). Mesoamerica was the most densely populated region in the Americas during this period, with between six and twenty-five million people (McEvedy and Jones 1978). When Columbus arrived, the Mayan civilization had collapsed and probably numbered about two million (*Encyclopedia Britannica*). Estimates for the entire continent vary widely, but figures of between thirty and fifty million people represent the mid-range. Michael Crawford surveyed the work of numerous other authors before deriving the regional estimates shown below (Crawford 1998). His estimate for North America is considerably lower than Ramenofsky's and Thornton's estimates of between five and twelve million (Ramenofsky 1987, Thornton 1987).

North America	2 million
Central America	25 million
Caribbean	7 million
South America	10 million
Total	44 million

EUROPEAN CONQUEST

The first Europeans to reach the New World made virtually no impact on native populations. Vikings reached North America 500 years earlier than Columbus on exploratory expeditions from well-established colonies in Greenland (Ingstad 1985). Originally, the documentation for these voyages was limited to rather vague Icelandic literature recording "Erik the Red" Thorvaldsson's trips to a place called Vinland in AD 986 (Wahlgren 1986). However, archaeological digs have since uncovered a Norse settlement on the northern tip of Newfoundland at a place called L'Anse aux Meadows (Figure 6.1). Norse artifacts and evidence of iron smelting, typical of Norse settlements in Norway, Iceland, and Greenland, were found here. A series of radiocarbon dates fixed the time of occupation to about AD 1000, corresponding surprisingly well with the written accounts. It is unlikely that these people ever established a long-term colony. They probably deserted the outpost soon after they established it and were probably either exterminated or assimilated by local inhabitants in contrast to the outcome of European contact centuries later, farther to the south.

The arrival of Columbus ushered in centuries of catastrophe and the collapse of the native populations throughout the length and breadth of the Americas. Eurasian societies enjoyed considerable advantages over Native American societies in food production, technology (particularly weapons), social organization, and written communication. The most potent weapon that Europeans brought with them was disease, which occupied center stage in the devastation that accompanied colonization.

The lack of resistance among America's natives to European diseases was among the most crucial factors that tipped the balance in favor of the invading force. For many centuries, Eurasian cultures had depended on a large number of domestic animals. Most of the microbes responsible for these infectious diseases had come from animals and had evolved over thousands of years, as did the European immunity and genetic resistance to them (Diamond 1997). The extensive trade networks throughout Eurasia, which facilitated the spread of technology such as metallurgy, the wheel, and agriculture, had ensured the intercontinental spread of disease and the simultaneous build-up of resistance.

Europeans depended on animals for protein (meat and milk), wool, and hides. More importantly, domesticated animals provided a major mode of land transport for people and goods, which proved to be a boon for crop production both in terms of pulling ploughs and in providing manure. And animals were invaluable in warfare. Among the New World domestic animals, none of them could transport people or pull a cart, plough, or chariot, and they contributed little to the diet of Native Americans. Had it not been for the post-glacial extinction of the large mammal species across North and South America, history may have taken a very different course after 1492. Disease and other factors associated with colonization, such as warfare, repatriation, and alcoholism, decimated the native populations. Before a century had passed since the Europeans' arrival, the population of Mexico and the Caribbean are thought to have declined from about 25 million to approximately one million. Similar tragic levels of mortality were common in native populations throughout the Americas (Crawford 1998).

Over the ensuing centuries, large numbers of Europeans and Africans made their way to the Americas where they intermarried with Native Americans. The highest level of intermarriage has occurred among Indian tribes in North America (Crawford 1998), less in the Southwest than in other areas of North America. By comparison, the rates of intermarriage in Latin America are much lower except in the areas adjacent to the great Aztec and Andean civilizations; and in Argentina, Uruguay, and Chile, where European settlement has been the most dense, only remnants remain of the former Indian populations (Crawford 1998).

7

Native American Molecular Genealogies

Should some investigator find new methods to pursue research on the "blood line" of a particular individual, family or people, he or she might find that some Native Americans are directly descended from Nephites of ancient times, that some are descended in part from others in Lehi's or Mulek's parties, that some are of Jaredite origin, and that still others have no discernible connection to any of those. Scientific, genealogical, or historical methods are not available; but more important, the scriptures indicate that the results would not matter as far as the Church and the gospel are concerned.

—John Sorenson, 1985

Like detectives arriving on the scene of a thousand-year-old crime, scholars searching for the origins of Native Americans have had to be content with sifting for clues through decaying buildings and the debris of prehistoric cultures. Yet the results of this archaeological research provide the backbone of historical thought concerning the New World (Collingwood 1946). In spite of the difficulty of the task, a relatively clear picture has emerged regarding when the ancestors of Native Americans arrived in the New World and where they came from. Like the most robust of scientific theories, it gains much of its strength from the fact that research across multiple disciplines points to the same conclusion. Molecular biology is simply the latest technology to be taken up by anthropologists—perhaps the most significant since radiocarbon dating—and makes an important contribution to the understanding of American prehistory.

Just as the surname Yamamoto reveals the country of its origin, molecular surnames and maiden names written into mitochondrial and Y chromosome DNA often disclose geographical and genealogical ties. But how useful are these molecular lineages when they reveal nothing about fourteen out of sixteen great-great-grandparents? In isolation, a person's mitochondrial or Y chromosome DNA lineage tells us little about his or her individual ancestry since they only represent a single maternal or paternal line. The real power of DNA lineage analysis is revealed when it is used to study human populations. If sufficient numbers of individuals are sampled from a population, the DNA lineages from a group collectively reveal a great deal of information about the ancestry of the population. The maternal and paternal DNA genealogies of Native Americans provide information of remarkable clarity and enable New World pedigrees to be constructed that span many thousands of years. In agreement with anthropological and archaeological research, the molecular pedigrees of Native Americans cluster on the Asian branch of the human family tree.

NEW WORLD FEMALE GENEALOGIES

Douglas Wallace at Emory University in Georgia was among the first to look closely at Native American mitochondrial DNA. In 1985 he and his research staff at the Center for Molecular Medicine found that Native Americans carried a subset of the maternal lineages found in Asia. A crucial milestone in this research was published in 1990 by Theodore Schurr, a student on Wallace's team (Schurr et al. 1990). Schurr found that native individuals in three well-separated populations—the Ticuna from western Brazil, the Maya of the Yucatan Peninsula, and the Pima from Arizona—shared DNA lineages that fell into four distinct groups. These four lineage families are now known by the rather nondescript labels A, B, C, and D (Torroni et al. 1992; see appendix A for a more detailed description of New World mitochondrial DNA lineages). When compared to maternal lineages from several other world populations, all four were found at moderate frequencies in Asian populations but were absent in all others. While it was clear that Native Americans carried Asian DNA lineages, it was also clear that not all Asian lineages reached the New World.

In the decade following the publication of Schurr's research, numerous other Native American and Asian populations were surveyed by several research teams. Presently, the maternal DNA lineages of over 7,000 Native Americans have been determined from about 175 native groups (see appendix B) scattered from the steamy jungles of Central America and the Amazon to the icy extremities of Alaska, Greenland, and Tierra del Fuego. This research has confirmed that the A, B, C, and D maternal lines are the major female "founding" lineage groups among Native Americans (Schurr 2000). These lineages were present among the ancestors of Native Americans who walked to the New World from Asia. About 96.5 percent of Native Americans have mitochondrial DNA belonging to one of these four founding lines, although populations differ greatly in the proportion of each lineage they contain.

In several populations in Alaska, Canada, and northeastern North America, a significant proportion of individuals have female DNA lineages that are distantly related to the four main founding groups. Originally it was thought that these new lineages might have been introduced after the arrival of Columbus, but when researchers looked closer, they found that most of them came from a fifth, though minor, founding line now known as the X lineage. It occurs at the highest frequency (about 25%) in several tribes in the northeastern part of North America (Brown et al. 1998) and at lower frequencies throughout the remainder of North America (Smith et al. 1999). Of the approximately 7,000 Native Americans who have been tested, about 2 percent have mitochondrial DNA belonging to the X lineage family. The frequency of the five female lineages among Native Americans from the continents of North and South America are shown in the following table.

Table 7.1 Maternal DNA lineages among Native Americans from North and South America.

	Maternal Lineage (number of individuals)						
Population	A	B	C	D	X†	Other	Total
Alaska	288	4	13	379	0	24	708
Greenland	82	0	0	0	0	0	82
Canada	443	42	82	29	55	3	654
United States*	554	633	379	185	61	16	1,828
Central America**	291	117	77	22	0	4	511
South America	676	1,175	914	683	0	86	3,534
Total	2,334	1,971	1,465	1,298	116	133	7,317
Percentage	32%	27%	20%	18%	2%	2%	100%

Source: See appendix B for a more detailed description of the occurrence of these lineages in New World populations and a listing of source references. †X lineage numbers are estimates based on Table 2 in appendix B. *This category refers to the continental United States. **Central America includes the Caribbean.

Unlike the A, B, C, and D lineage families, X occurs at frequencies of about 4 percent in western Eurasian populations. It is particularly common in the Middle East, where it occurs at a frequency of about 25 percent in some populations. The X lineage was recently discovered in Central Asia, where it occurs in about 3.5 percent of the Altaian populations of southern Siberia (Derenko et al. 2001). The virtual absence of the X lineage in eastern and northern Asia raised the possibility that the X lineage had come to the Americas by way of Europe, but the most likely route of travel would have been easterly through Asia. However, a westerly route across a frozen Atlantic could not be ruled out. Recent work on the Native American X lineages has shown that they are distantly related to both the European and Altaian X lineages (Brown et al. 1998, Reidla et al. 2003). In a global study of the X lineage family tree, it is clear that the Altaian lineages—those studied thus far—are not directly ancestral to the New World lineages. Meanwhile, the discovery of X mitochondrial DNA in ancient samples from 700 to 7,000 years old confirms that the lineage arrived in America with the original founding population (Hauswirth et al. 1994, Stone and Stoneking 1998).

About one in every 200 Native Americans has a maternal DNA lineage that is common in Europe or Africa and is not found in Asia. In a survey of about 800 individuals from almost fifty tribes in North America, seven individuals were found to have European lineages including six H lineages and a single T lineage (Lorenz and Smith 1996). One individual in the Mexican Mayan sample (Torroni et al. 1992) and a small number of the Canadian Ojibwa (Scozzari et al. 1997) also belonged to lineage H. The H and T lines are two of the most common female lineage groups among Caucasians and occur at a combined frequency of between 60 and 80 percent in European populations (see chapter 9). Spain has the highest proportion of these two female lineages. Several Native Americans have been found whose mitochondrial DNA belongs to the L lineage family, which is found in about 76 percent of Africans. This includes individuals from the Seminoles of Florida (Huoponen et al. 1997) and among the Mixtec and Zapotec Indians of Mexico (Torroni et al. 1994a, Lell et al. 1997). Given the substantial immigration of Europeans and Africans into the Americas during the last 500 years, it is not surprising that such lineages would be found among Native Americans. The occurrence of maternal lines that are common in Spain is consistent with the prominent role the Spanish played in the early discovery and colonization of North and South America.

The age of Native American maternal lineages gives a molecular clue for when the first people entered the unoccupied New World. Each of the lineage groups found among Native Americans (A, B, C,

D, X) contains lineages with DNA sequence changes that occurred after the original founders became separated from their Asian source population. Since the approximate rate at which mitochondrial DNA accumulates new mutations is known, it is possible to estimate the age of ancestral lines in a technique called coalescence analysis.

The coalescence times for the five founding New World lineages —in other words, the approximate date when the Asian and American populations separated—average between 20,000 and 40,000 years ago (Brown et al. 1998, Lell et al. 1997, Bonatto and Salzano 1997, Forster et al. 1996, Torroni et al. 1994b). Coalescence times for the Siberian A, B, C, and D lineages are similar (Starikovskaya et al. 1998). DNA variation found in the X lineages on either side of the Bering Strait is sufficient to indicate that they have been separated for between 30,000 and 40,000 years and that American X lineages had a common ancestor about 18,000 years ago (Reidla et al. 2003). Overall, the mitochondrial DNA evidence supports the idea that the five founding female lineages may have been brought into America before the height of the last glacial encroachment which occurred about 18,000 years ago.

NEW WORLD MALE GENEALOGIES

Research on female genealogies progressed ahead of the genealogies of New World men. Initially it was a daunting task to try to read the molecular surnames on the Y chromosome, but over time, research has uncovered an enormous amount of genealogical information hidden on this chromosome. Early reports on Native American paternal genealogies were complicated by the fact that few research teams were using the same genetic markers or nomenclature for the lineages they were reporting. The research did find, however, that most Native American males have Y chromosome DNA lineages that are relatively rare in Asia and generally absent in Europe (Karafet et al. 1999).

In 2002 an international consortium of scientists constructed a global Y chromosome lineage tree and introduced standardized nomenclature to avoid confusion and improve communication among scientists (YCC 2002). The most prevalent Y chromosome lineage group among Native Americans is now known to be the Q lineage. It occurs in over 90 percent of South American Indians (Bortolini et al. 2003) and about 75 percent of Indians from North America (Lell et al. 1997, Zegura et al. 2004, Karafet et al. 1997, Pena et al. 1995, Santos et al. 1996, Santos et al. 1999, Underhill et al. 1996). Lineage Q occurs at a moderate frequency in Asian populations but is rare in Europe or the Middle East. Native American Q lineages are clearly descended from Asian Q lineages because they share common DNA markers (Zegura et al. 2004). Two other lineages have been found in native populations that are believed to have been present with the original founders. These are lineages C and P, which appear to be more common in North America than in South America (Bortolini et al. 2003, Zegura et al. 2004).

Other lineages of either European or African origin are most common in North America, where they occur at a frequency of about 10 percent compared to about 4 percent in South America. The higher occurrence of European lineages among Native American men is not surprising given that the front line of the European colonizers consisted mostly of males.

In contrast to the age estimates for maternal lineages, several recent reports on the age of New World paternal lineages suggest a late entry into the Americas. Estimates for the age of both the C and Q lineages in Native Americans range from 10,000 to 17,000 years before the present (Zegura et al. 2004). The estimates for a common mutation found among Q lineages, which occurs at a low frequency in Asian Q lineages, are about 14,000 years before the present (Bortolini et al. 2003, Seielstad et al. 2003), indicating a much more recent arrival than the 20,000- to 40,000-year estimate calculated on the basis of mitochondrial DNA variation. The Y-based research is compatible with the ages of the earliest archaeological sites in the Americas, dated at approximately 14,000 years old. The discrepancy may be due to a miscalculation of the rate of mitochondrial DNA mutation. It is most likely due to a false assumption about the number of founding DNA lineages. Where it has been assumed that there were only five founding lineages, this may not have been the case. It is, in fact, quite likely that there were more than one line in each of the five founding groups. Both of these factors would lead to overestimates of the age of the maternal lines (Bortolini et al. 2003). Recently, looking at a larger portion of the mitochondrial DNA, researchers needed to revise their estimates for the differentiation of the A, B, and C lineages—moving them closer to the Y chromosome estimates—to about 21,000 years ago (Silva et al. 2002).

SIBERIAN HOMELANDS

To predict where in Asia one might find the source population for the American founding population requires cooperation by scientists from different disciplines. For instance, one could test the frequency of New World lineage groups among Asian people, but since Asian populations have not always been static, one has to rely on archaeologists and anthropologists to appreciate the history of the potential source populations.

Asian populations that have the highest proportions of New World maternal lineages are generally located in central Asia near southern Siberia, northern Mongolia, and China. Populations that have all four founding lineages—Mongolians, Altaians, Buryat, and Tuva—occur in the vicinity of Lake Baikal, a large freshwater lake in southern Siberia (Figure 7.1). The only exception is the Tibetans, who are located in southern Asia. Other genetic evidence suggests how closely related Tibetans are to northern Asian and Siberian populations, which can be explained by the geography, Tibet being more accessible from the north than from the southeast (Torroni et al. 1994c). Tracing the genetic trail of Native American paternal lineages brings one to a similar region in Asia (Karafet et al. 1999) among the Ket, Selkup, and Altaian populations of the area west of Lake Baikal (Figure 7.1).

The Lake Baikal region is the ancestral homeland of northeastern

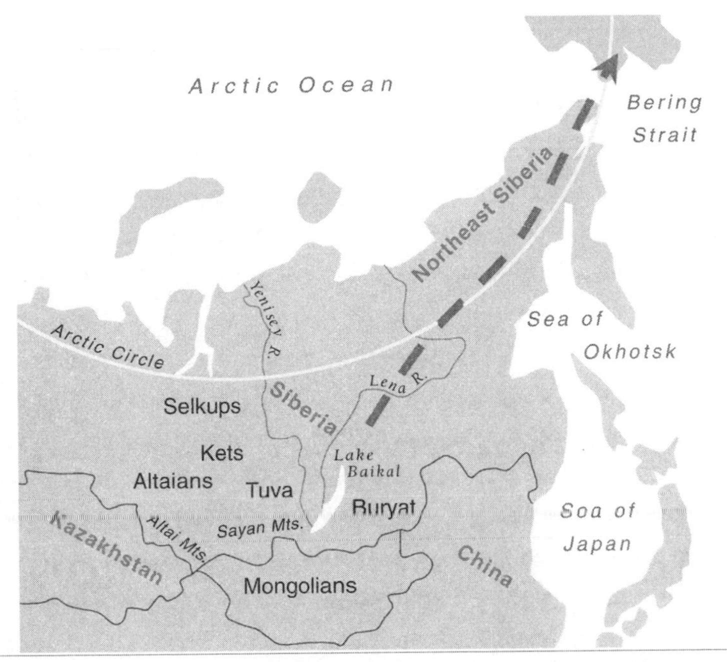

Figure 7.1 Asian populations with the highest proportion of New World maternal and paternal lineages are located in the vicinity of Lake Baikal in southern Siberia. The Altaians, native to the Altai Republic (South Siberia) and numbering about 60,000, appear to share the strongest genealogical ties with Native Americans. Asian archaeological research indicates that the ancestors of these groups once lived around the shores of Lake Baikal. The dashed arrow represents the likely route of migration to North America.

Siberian populations and of the people who inhabit the region between the Sea of Okhotsk and the Yenisey River. While the Kets and Selkups currently inhabit western Siberia and the Yenisey River Valley, their ancestors originated farther south on the slopes of the Sayan and Altai Mountains (Popov and Dolgikh 1964, Prokof'yeva 1964). The paternal and maternal ancestors of the Native Americans likely derive from the region encompassing the Altai and Sayan Mountains and the region immediately surrounding Lake Baikal (Figure 7.1).

During the last Ice Age, northeastern Siberia was largely uninhabitable. Cultures living in southern Siberia west of the Yenisey River shared stone implements and cultural traditions similar to those of European groups in the west, with whom they had considerable contact. The extinction of large grass-eating animals, such as the mammoth and rhinoceros, at the end of the Ice Age drove some Siberian tribes to adopt a more nomadic existence, hunting smaller animals such as horses and reindeer over the thawing expanse of northeastern Siberia. These nomadic cultures began to diverge from those found in the west and Lake Baikal region as they migrated northeastward toward Beringia (Okladnikov 1964).

Anthropologists also believe that from at least 10,000 years ago, the Lake Baikal region was populated by mixed tribes with Caucasoid and Mongoloid anthropological features. These tribes were displaced by Mongoloid groups of central Asian origin (Alexeev and Gokhman 1984). The occurrence of the rare X lineage among Altaians provides additional evidence that the ancestors of this population may have been the source of the original founders who migrated to the New World. Recent mixture with Europeans cannot account for the occurrence of the X lineage among Altaians since the lineage is extremely rare among contemporary Siberians with recent European ancestry. Hence, the X lineage likely came from the gene pool of the ancestors of the Altaian people, indicative of an early influence from Europe.

SIBERIAN COMPANIONS

DNA genealogy is also now shedding light on the ancestry of some of man's closest companions who likely accompanied him into the New World. For instance, dogs are the only domesticated species found across both Eurasia and the Americas before the transoceanic voyages of the fifteenth century. It is known that dogs were domesticated from gray wolves (*Canis lupus*), but this species has a wide distribution on either side of the Bering Strait (Savolainen et al. 2002). Molecular research has helped elucidate where the wild ancestors of American dogs are likely to have lived.

Mitochondrial studies suggest that New World dogs were faithful companions during the early settlement of the Americas. DNA sequences obtained from pre-Columbian dog remains in Latin America and Alaska have been compared to the DNA of dogs and wolves from around the world (Leonard et al. 2002), and it is now clear that New World dogs occupy the same branch of the canine family tree as modern Eurasian dogs. The American gray wolf is a distant cousin. It would appear that when humans crossed the Beringian land bridge into the Americas, they were accompanied by their canine companions.

Further evidence for Native American ancestry can be found by tracing the organisms that have plagued humans for thousands of years. *Helicobacter pylori,* a chronic gastric pathogen of humans, is found virtually worldwide and can be classified according to its geographical distribution. Analysis of the bacterium within native human populations shows that the East Asian strains occur widely in the New World, indicating previous human contact between East Asians and Native Americans.

A fascinating piece of evidence that points to southern Siberia as the source of New World populations comes from the study of the distribution of a human T-cell virus (Neel et al. 1994). This generally harmless virus is a freeloader, passed on in breast milk from generation to generation by mothers feeding their babies. Consequently, its pattern of inheritance resembles that of mitochondrial DNA in that it is passed from mother to offspring. A particular strain of T-cell virus (HTLV-II) common in Native Americans has not been found in northern or eastern Siberians. However, it exists in the people who live in the regions of Mongolia, Manchuria, and southern Siberia.

NEW LIGHT FROM OLD DNA

The capacity of mitochondrial analysis to reveal maternal lineages in DNA applies to ancient remains as well and allows scientists to consider the relatedness of past and present native populations. This promises to shed even more light on the early settlement of the Americas. Among the most controversial questions scientists currently grapple with is why the earliest human skulls found in the Americas are morphologically more similar to contemporary Europeans than Native Americans. When the skull of Kennewick Man was found in the state of Washington, scientists were eager to determine the DNA lineage of the remains. Unfortunately, repeated attempts by researchers from the University of California at Davis, the University of Michigan, and Yale University have failed to detect ancient DNA in the remains. Most likely, environmental conditions at the site were not conducive to the preservation of DNA. Despite this failure, scientists remain optimistic that some of the earliest skulls will contain DNA that is suitable for analysis.

In a well-publicized 1999 BBC documentary, "Ancient Voices," some seemingly compelling evidence was presented that the first Americans may have been Australians. This is the view of a few scholars who see in the morphology of early skulls found in South America more similarities to Melanesians and Australians (Neves and Pucciarelli 1998) than to Siberians. This led to speculation about marathon ocean voyages by Australians and their subsequent massacre at the hands of invaders from Asia. The only remaining traces of these earliest inhabitants were claimed to have been found in the morphology of the Native Americans of Tierra del Fuego, islands at the remote southernmost tip of South America. The truth is likely to be much simpler. Ancient DNA has been isolated from teeth and bones of sixty Tierra del Fuegians and all were found to have either a C or D mitochondrial DNA lineage, commonly found in contemporary Native Americans (Lalueza et al. 1997). The absence of A and B lineages may indicate why their facial features are different. It is probable that the early colonizers of the Americas had a range of genetic backgrounds, resulting in morphological variation in populations in different parts of the two continents. Perhaps the Fuegians carried a discrete subset of that genetic variation that resulted in their distinctive morphology. It is also possible that the X maternal lineage, which is common in North America and has roots in Europe, may be a remnant of an early migratory group with slightly more European or African appearance.

DNA analysis of ancient remains from the Ohio Valley has driven a belated nail into the coffin of the Mound Builder myth and its supposed white inhabitants. The people of the so-called "enlightened" race that was responsible for the mounds in the Ohio and Mississippi Valleys are now known to be Native Americans belonging to the Adena and Hopewell cultures. Maternal DNA lineages have been determined for ninety-seven skeletal samples obtained from two Adena mounds in Kentucky and Hopewell mounds in Ohio and Illinois (Bolnick 2003, Mills 2003). All of the maternal lineages belong to one of the five founding lines common to contemporary Native Americans (see appendix B). The DNA analysis confirms that the idea of a superior race of Mound Builders was pure fantasy.

CONCLUSION

The DNA research supports the morphological evidence of a close relationship between Native Americans and Mongoloid peoples from Asia. The reason for this is that human morphology is largely genetically predetermined by DNA. Many of the very slight differences in the DNA of genes controlling these traits, as superficial as they may be, will doubtless be shared between populations. While the maternal and paternal DNA lineages do not directly measure the differences in these genes, their presence in both groups of people indicates the sharing of many closely related genes contained somewhere on the chromosomes. Molecular research may yet shed more light on the intriguing observation of Caucasian features in the earliest skulls of indigenous Americans. The occurrence of the X lineage among the Altaians is perhaps a sign of a Caucasian influence among the most likely ancestors of America's founders.

The geological and environmental history of the lands joining the Old and New Worlds strongly suggests that entry into North America from Alaska/Beringia would have been virtually impossible between about 14,000 and 25,000 years ago. There appear to be two windows of opportunity when it would have been practicable for humans to enter the Americas. The first was before the Ice Age began 25,000 years ago. The second brief opportunity came after the Ice Age and before the flooding of Beringia sometime later than 14,000 years ago. When the DNA evidence is considered in the context of the geological and archaeological history, it would appear most likely that the first Americans reached the continent after the last glacial maximum which occurred about 18,000 years ago. The Y-chromosome research, in particular, supports the view that America was first colonized as recently as about 14,500 years ago, followed by a rapid population dispersal throughout both continents. Waves of Asian migration were largely impeded about 10,000 years ago by the flooding of Beringia. The Eskimos of Siberia and North America appear to be remnants of a population that inhabited Beringia at the conclusion of the last Ice Age and who became separated from their relatives as the sea rose (Starikovskaya et al. 1998). Despite the rise in the ocean, the icy waters of the Bering Strait never have been an effective barrier to Eskimo cross-migration, which has occurred frequently between the two continents over the years.

8

Polynesian Molecular Genealogies

It is extraordinary that the same nation should have spread themselves over all the isles in this vast Ocean from New Zealand to this Island which is almost a fourth part of the circumference of the Globe.

—Captain James Cook, Easter Island, 1774

James Cook sailed into Polynesia in 1769 aboard the converted coal ship *Endeavour*. At the wishes of the Royal Society, he charted a course to Tahiti to record the transit of Venus on 3 June. In the wake of the Spanish, Portuguese, and Dutch naval explorers of earlier centuries and equipped with a telescope, sextant, and the newly invented chronometer, Cook sailed at the peak of European exploration of the Pacific. Unlike Columbus, he knew exactly where he was, which is plainly revealed in the numerous charts he drew that retain their accuracy to this day. After completing his assigned task, he opened a set of orders with instructions to claim a great southern continent in the name of the king of Great Britain (Taylor 1990). On this and his succeeding voyages, Cook proved himself one of the great maritime explorers of history, playing a significant role in charting and discovering the islands of the Pacific. A skillful and observant sailor, his statements on the common ancestry of Polynesians remain as true today as the day he uttered them.

The largest islands in the Pacific region are Papua New Guinea and the continent of Australia at the western extremities. Until the end of the last Ice Age about 11,000 years ago, lowered sea levels exposed a continuous land mass named Sahul, which encompassed Papua New Guinea, the Australian mainland, and Tasmania. The shallow 100-mile-wide Torres Strait now separates Australia from New Guinea. Moving east into the Pacific, one encounters the geologically modern islands scattered throughout the regions of Melanesia, Micronesia, and Polynesia. The latter encompasses a vast sector of the Pacific—twice the area of the United States—situated between the Hawaiian Islands, Easter Island, and New Zealand. The waters of the Pacific Ocean cover approximately 98 percent of this region.

COLONIZATION OF THE PACIFIC

Humans entered the Pacific much earlier than the Americas. Archaeological and linguistic research leads scientists to believe that there were multiple waves of Pacific settlement. The first phase resulted in the colonization of Sahul, culminating in the settlement of Australia between 50,000 and 60,000 years ago (Thorne 1980, Roberts et al. 1990, Thorne et al. 1999). By 30,000 years ago, the original settlers had spread throughout most of the Australian continent. However, the predominantly arid environment of Australia limited indigenous population growth. When Tasmania became separated from the mainland about 13,000 years ago at the conclusion of the last Ice Age, the aborigines who lived there were effectively isolated from the mainland populations. More recent waves of immigration from Papua New Guinea probably occurred as recently as 6,000 years ago. One of these groups probably brought domestic dogs with them, which dispersed into the wild on mainland Australia. These wild dogs, known as dingos, have largely interbred with domestic dogs belonging to Europeans.

The founding colonizers of Papua New Guinea were displaced by successive waves of people from neighboring islands to the west who spread to the adjacent Melanesian islands in the Bismarck Archipelago and the Northern Solomons by 6,000-15,000 years ago (Bellwood 1987, Wurm 1967). The settlers of western Micronesia are believed to have originated in Indonesia or the Philippines (Bellwood 1979), while eastern Micronesia appears to have been settled from Melanesia, possibly Fiji (Spriggs 1985).

Settlement of the remainder of Melanesia, parts of Micronesia, and much of Polynesia was achieved by people who spoke an Austronesian language (Spriggs 1985). The linguistic ties between populations in this region provide strong evidence for a close genealogical link (Figure 8.1). The last phase of Pacific island settlement was rapid (Diamond 1988), taking about 2,000 years until complete in about AD 1000 (Table 8.1). This expansion is linked to the spread of distinctively styled Lapita pottery, named after an archaeological site in New Caledonia where it was first described (Bellwood 1979, Spriggs 1985). Lapita pottery, with its striking geometric designs, has been found all the way from coastal Papua New Guinea eastward to Samoa (Bellwood 1979).

The Polynesians are descended from the Lapita people, who were highly mobile sea colonists and skilled agriculturalists. Their rapid expansion from Melanesia to Polynesia was probably made possible by the development of exceptional skills of navigation and the invention of the outrigger, which brought stability to ocean-going canoes (Bellwood 1979). Colonization of new islands was unlikely to have been the result of disoriented fishing voyages because women, plants, and animals were carried on the canoes. These voyages were probably never comfortable or dull, as domesticated pigs, dogs, and fowl accompanied these seafarers across the Pacific. Rats and lizards from Asia also found their way into the Pacific aboard Polynesian sailing craft. The rodents and reptiles may have been taken as a food source, but they probably undertook most of these migrations as stowaways.

Crop production in the Pacific was typically based on the shifting cultivation of a number of tubers and fruits, together with an intensive use of marine resources (Bellwood 1987). Cereals, while important in Southeast Asia, were not grown on the Pacific Islands. The major cultivated plant foods included taro, yam, sweet potato, coconut, breadfruit, banana, and the less common pandanus, gourd, and Tahitian chestnut. All of these plants were domesticated in Southeast Asia with

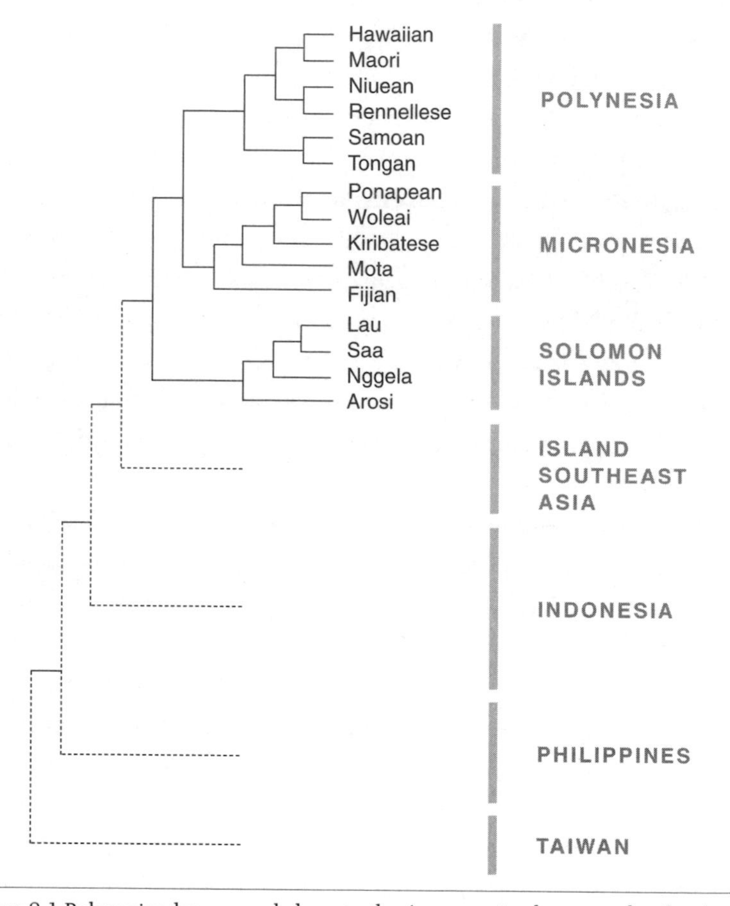

Figure 8.1 Polynesian languages belong to the Austronesian language family, the world's largest. Word data from Austronesian languages from Southeast Asia and Oceania have been used to construct tree-like arrangements of related languages. When closely related languages are positioned near each other, we see clear paths of language evolution in the Pacific trailing back through Southeast Asia to Taiwan. Adapted from Gray and Jordan 2000.

the notable exception of the sweet potato (*Ipomoea batatas*), which is almost certainly an introduction from South America, based on botanical relatedness (Solheim 1970). Peter Bellwood from the Australian National University, Canberra, argues that it was probably introduced into Polynesian culture about 1,000-1,500 years ago before the Polynesian expansion to the extremities of Easter Island and New Zealand, where it became an important crop (Bellwood 1987). In the United States, some varieties of sweet potato are incorrectly called yams, an entirely different plant belonging to the genus *Dioscorea*.

New Zealand, Hawaii, and Easter Island lie at the ends of the chain of Polynesian migration. Maori oral tradition recounts epic voyages requiring great navigational skills. The most widely accepted theory of New Zealand colonization is a planned settlement with a significant number of colonists from eastern Polynesia traveling thousands of kilometers with their animals, plants, and cultural artifacts (Murray-

Table 8.1 Approximate date of contact or colonization of the South Pacific Islands by the ancestors of the Polynesians

Island	Date
Contact	
Philippines	2000 BC
Coastal Papua New Guinea	1000 BC
Colonization	
Fiji	1000 BC
Samoa and Tonga	1000 BC
Marquesas	AD 800*
Easter Island	AD 800
Hawaii	AD 800
Society Islands	AD 800
Cook Islands	AD 800
Aotearoa (New Zealand)	AD 1200

* Possible site of introduction of the sweet potato into Polynesia.

Sources: Spriggs 1996, Bellwood 1979, Bellwood personal communication.

McIntosh et al. 1998). It has been proposed that the settlers arrived over a period of several generations beginning over 800 years ago. Approximately 500 individuals would have been required to provide the number necessary to occupy the early coastal sites found throughout the country and to import the range of skills, traditions, and knowledge essential for successful colonization (McGlone et al. 1994).

When the Polynesians first arrived in New Zealand, they were almost certainly accompanied by domesticated pigs and chickens, as these animals were a major source of protein that sustained other Polynesian settlements throughout the Pacific. The immigrants soon found that their new home had a plentiful supply of large birds, seals, fish, and shellfish capable of sustaining a flourishing population. Surely the most arresting animals to greet the first Maoris were the large flightless birds which they named "moa," the Polynesian word for chicken. Some of these "chickens" stood over three meters (ten feet) tall and weighed in at up to 250 kilograms (550 pounds). Not surprisingly, the Maoris soon lost interest in their domesticated chickens and began harvesting the moa, which were slow moving and lacked any fear of humans (Flannery 1994). Numerous moa harvesting sites have been discovered throughout New Zealand, particularly on the South Island, and the large numbers of moa eaten at these sites sustained very large populations. Within 400 years of the arrival of the Maori, all species of moa were extinct, and by the time Europeans arrived, the Maori were suffering from severe food shortages. The closest surviving relative of the moa is the flightless kiwi, essentially defenseless birds whose evolution was due to the unique ecological history of New Zealand. Mammals did not exist in New Zealand, or throughout most of the Pacific, until the arrival of man. Bird species in New Zealand evolved to fill a wide range of niches in the absence of mammal predators.

The precise origin of the Polynesian people who carried the Lapita culture to the Pacific is unknown and, consequently, it is among the most debated questions in Pacific archaeology. Most of the evidence points to Taiwan and Indonesia for the primary source populations.

Peter Bellwood believes that the Polynesians originally migrated from China or Taiwan. Their migration took them steadily through the islands of Southeast Asia and the heavily populated areas of Melanesia. Through Melanesia, they island-hopped between coastal settlements, then radiated out into the Pacific (Bellwood 1979). They are likely to have mixed considerably with local people en route. Some infer that Maoris arose first in Melanesia, gradually spreading into the islands of Southeast Asia, Melanesia, and then Polynesia (Terrell 1986). A third hypothesis advanced by Thor Heyerdahl considers South America to be the ancestral home of the Polynesians (Heyerdahl 1950).

POLYNESIAN DNA GENEALOGIES

As with the Americas, the living inhabitants of the Pacific trace their molecular roots back to Asia. Mark Hertzberg's research team at the University of Sydney (Hertzberg et al. 1989) published the first survey of the mitochondrial DNA lineages of Polynesians in 1989. They found that most of the people they surveyed (90%) from the islands of Samoa, New Zealand, Niue, the Cook Islands, and Tonga had mitochondrial DNA with a short, 9-base deletion commonly found among Southeast Asians. This deletion is one of the characteristics of lineage B, which is common in southern Asia and is one of the four major lineages among Native Americans. They examined other South Pacific populations, as have additional researchers since then. In most areas of Polynesia, more than 95 percent of the native population has a type B maternal DNA lineage. The general trend is one of a steady increase in the frequency of the B maternal lineage as one moves from Asia to the extremities of Polynesia. The B lineage is essentially absent among Papua New Guinea highlanders and Australian Aboriginals.

The occurrence of the B lineage in the Pacific Islands and the Americas raises the question of recent common ancestry. The Asian, rather than American, origin is supported by the occurrence of two unique mutations in Polynesian mitochondrial DNA. The Asian B lineage has a *C* at base 16,217, an *A* at base position 16,247, and a *C* at base 16,261—the *CAC* lineage. Polynesian B lineages always have a *T* instead of a *C* at base position 16,261—the *CAT* lineage—and almost always a *G* instead of an *A* at base position 16,247, giving rise to the *CGT* or Polynesian lineage. The *CAT* lineage is found in Southeast Asia and the *CGT* lineage has been found in coastal regions of Indonesia and Papua New Guinea. This strongly suggests that the people who brought the Polynesian lineage into the Pacific came from Asia (Figure 8.2). The *CAT* and Polynesian (*CGT*) lineages have not been found in American Indian populations (Bonatto et al. 1996, Sykes et al. 1995).

The accumulated research strongly supports a Polynesian female genealogy leading back exclusively to Asia. Lum and colleagues (1998) concluded from their research on mitochondrial and nuclear DNA that there is a close link between Southeast Asian and Oceanic populations. As one moves away from Asia through the Pacific Islands to the most distant islands of New Zealand, Hawaii, and Easter Island, the number of lineages decreases (Murray-McIntosh et al. 1998). Immigration from America would have increased the number of female lineages near the New World. Rather, the trend is consistent with a generally eastern migration of small founding groups that colonized the islands in succession, each new colonization resulting in a reduced number of maternal DNA lineages among those taking possession of the newly discovered island. The virtual fixation of the Polynesian lineage in the Maoris of New Zealand illustrates the endpoint of this trend.

The reduction in the number of female lineages does not mean that the colonizing parties were comprised of only a handful of individuals. Rosalind Murray-McIntosh of Massey University, New Zealand (Murray-McIntosh et al. 1998), believes the DNA evidence strongly suggests that there was a significant number of females in the canoes when the Polynesians colonized New Zealand. She estimates that between 50 and 100 women sailed in the party and that the evidence does not support a small founding group.

The discovery of the Polynesian lineage in the Malagasy (Figure 8.2) from the island of Madagascar (Soodyall et al. 1995) is a striking testament to the maritime skills of the Polynesians. Linguistic and

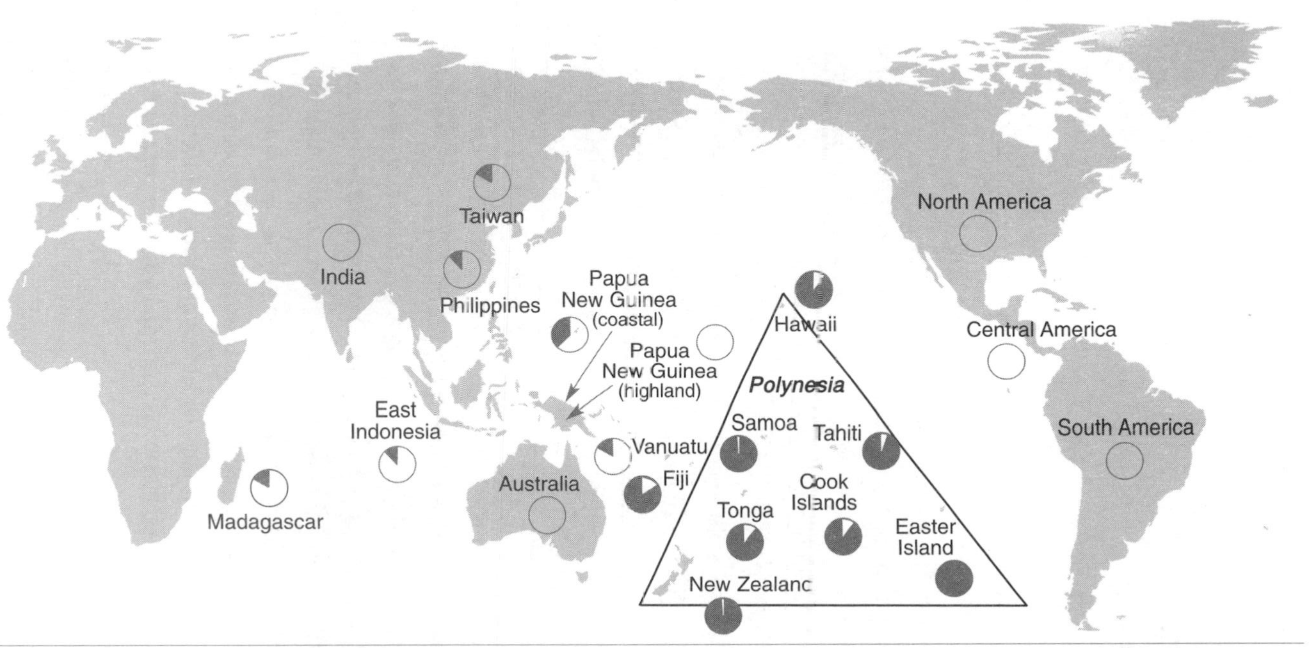

Figure 8.2 Distribution of Polynesian maternal B lineages (*CAT* and *CGT*) throughout the world. The occurrence of the ancestral *CAT* lineage in mainland and island Southeast Asia suggests the origin of the Polynesians there. The most widely accepted homeland for the ancestors of the Polynesians is Taiwan, but the descendants of the original Polynesians mixed considerably with other populations and spread most of the way around the globe. The triangle indicates the extent of Polynesia. Sources: Hertzberg et al. 1989, Lum et al. 1994, Lum and Cann 1998, Melton et al. 1995, Murray-McIntosh et al. 1998, Soodyall et al. 1995, Stoneking and Wilson 1989, Sykes et al. 1995, Torroni et al. 1994d.

archaeological evidence indicates that despite the fact that nearby Africans reached Madagascar, people from Indonesia got there first in about AD 400. At the time of these epic Polynesian voyages, few Europeans had ventured far beyond the Mediterranean. Again it would appear that these voyages were undertaken with a view toward colonization because women and probably family groups were among the migrating parties.

A closer examination of 655 Polynesian maternal lineages in 1995 revealed that about 6 percent do not belong to the B lineage family (Sykes et al. 1995). Most are so different that they probably arrived in Polynesia through contact with distantly related populations, probably those with whom the Polynesians mixed along their progressive journey out to sea. About 4 percent of the lineages among Polynesians originated with people from Vanuatu and Papua New Guinea. This is compatible with the colonization history of small groups of skilled seafarers who island-hopped through Southeast Asia and Melanesia, mixing with coastal populations. This is further supported by the discovery of the Polynesian (*CGT*) lineage among coastal peoples throughout Melanesia and Micronesia. The origin of the remaining 2 percent of Polynesian lineages is less clear. Some from this group are similar to Filipino lineages (0.6%) and others seem to be from western Europe, particularly from the United Kingdom. Two lineages found in single individuals from the Cook Islands and Tahiti matched Native American A and C lineages respectively. This may be the first genetic evidence of human contact between Polynesia and South America. An Asian origin for these lineages cannot be ruled out, as an identical A lineage has been found among the Japanese (Lum and Cann 2000).

Polynesian male genealogies are now sufficiently well understood to address the question of where the paternal ancestors of the Polynesians originated. Surprisingly, most of the men who migrated to Polynesia had Y chromosomes originating in Melanesia. The dominant Y chromosome found among Polynesian males belongs to the C lineage family. This Y lineage has been observed at frequencies of between 60 and 80 percent in Cook Islanders and Samoans (Hurles et al. 2002, Kayser et al. 2000). Outside of Polynesia, this particular C lineage has only been observed in Melanesia and eastern Indonesia (Table 8.2). A small proportion of Taiwanese Y chromosome DNA lineages occur among Polynesians (Hurles et al. 2002, Su et al. 2000), but most Polynesian males track as Melanesian. Other minor lineages observed among Melanesians and Polynesians include lineages O, K, M, and R (YCC 2002).

Based on the amount of variation present in Melanesian C lineages, it would appear that they were present in Melanesia for as long as 11,500 years, clearly predating the migration of Austronesian speakers into the Pacific (Kayser et al. 2000). Similar analysis of Polynesian C lineages suggests that Polynesians shared a common ancestor as recently as 2,200 years ago. The fact that some common Melanesian Y lineages are absent in Polynesia suggests that the original Polynesians were a small group that left several Y chromosome lineages in their wake as they sailed out into the Pacific.

The predominant Native American male lineage Q, found in

Table 8.2 Paternal (Y chromosome) lineages observed in Cook Islanders (bold) and their frequency in other populations of Melanesia, Asia, and Australia.

	Paternal Lineage (%)		
Population	C	O	other
Cook Islands	**82**	**7**	**11**
Papua New Guinea coastal	26	10	64
Papua New Guinea highland	3	0	97
Moluccas	15	12	73
Java	0	23	77
Han Chinese Taiwan	0	58	42
Taiwan aborigines	0	12	88
Australian aborigines	0	1	99

Source: Kayser et al. 2000. Lineages C and O are as defined in YCC 2002.

about 85 percent of New World males, is generally absent in the Pacific among Polynesians or Melanesians (Karafet et al. 1999, Hurles et al. 1998). The only place Native American Y lineages have been clearly identified is on the island of Rapa in eastern Polynesia (Hurles et al. 2003). Of sixteen distinct paternal lineages observed among Rapans, three were Native American and three were European. Further investigation into the history of Rapa revealed a dramatic population collapse on the island in 1864 due to smallpox or dysentery. Only twenty males are thought to have survived the epidemic. The source of the outbreak was traced to infected slaves on the illegal Peruvian slave ship *Cora,* which was overpowered by Rapans in 1863. While most of the crew were deported to Tahiti to face trial, five were assimilated into the Rapan population. Among these five men, three were either from Chile or Mexico and had Hispanic names (Hurles et al. 2003). This is the likely source for the European and Native American Y chromosomes found among the men of Rapa.

STAPLES AND STOWAWAYS

Further light has been shed on the ancestral movements of Polynesians by focusing on the animals and plants that accompanied them on their voyages. As noted earlier, rats were common passengers aboard prehistoric Polynesian canoes, either as food or as stowaways. The Pacific or Polynesian rat (*Rattus exulans*) is found throughout remote Oceania and as far into Southeast Asia as the Andaman Islands off the west coast of Thailand. DNA lineage studies of Pacific rat populations suggest that the deepest branches of the rat family tree occur in central Polynesia, in particular the Cook and Society Islands (Matisoo-Smith et al. 1998). In addition, the DNA evidence supports the view that there were multiple contacts between central Polynesia and outlying islands including Hawaii and New Zealand. One rather surprising observation is that rats from the Marquesas group of islands are not genetically diverse, implying that there were not repeated introductions of new rats onto these islands. Traditionally the Marquesas have been regarded as the primary center of human dispersal in east Polynesia, but the evidence strongly suggests that the Marquesas were isolated through much of the early period of Polynesian prehistory.

Similarly revealing genetic studies have been carried out on the small lizard, *Lipinia noctua,* which is found in the Papua region in New Guinea and throughout Oceania as far as the Hawaiian Islands in the northeast and Easter Island and Pitcairn Island in the southeast (Austin 1999). This lizard, commonly known as the moth skink, lives in close proximity to humans and likely stowed away on Polynesian canoes. Moth skinks living in New Guinea are genetically diverse; however, all central and eastern Pacific populations are genetically similar. Greater genetic diversity in a particular geographic location implies longevity. The limited diversity in Oceania supports the view that colonization of the far reaches of the Pacific occurred recently and was preceded by a rapid migration out of Southeast Asia.

DNA research is just now beginning to shed light on one of the most intriguing puzzles in Pacific history—how, when, and from where the sweet potato made its entrance into the Pacific. The earliest Spanish and Portuguese explorers of Polynesia, Micronesia, and Melanesia repeatedly observed *batatas,* or sweet potato, among the native crops (Sauer 1993). It was reported in such widely scattered places as the Marquesas, the Solomon Islands, New Zealand, and Guam, and there is firm evidence that the sweet potato was an important Polynesian crop at the time, deeply embedded in tradition and ritual. Sweet potatoes were a staple crop in the highlands of New Guinea as early as the seventeenth century. Carbonized, fossil sweet potato tubers have been found in New Zealand dating to about 600 years ago.

Despite being one of the most important agricultural crops in the world, the domestication history of the sweet potato is poorly understood. It was widely grown in Mesoamerica and the Andes, and there is evidence that it was first domesticated as early as 10,000 years ago. It is generally accepted that the sweet potato was domesticated somewhere between the Yucatan Peninsula in Mexico and the Orinoco River in

Venezuela. Since 1985 the International Potato Center (CIP) in Lima, Peru, has been gathering extensive collections of sweet potato cultivars from throughout the Americas and the Pacific. DNA fingerprinting studies reveal that most sweet potato genetic diversity occurs in Mesoamerica, while there is limited diversity in Peru (Zhang et al. 2000). Sweet potato cultivars grown in New Guinea and throughout Oceania have broad genetic diversity comparable to Mesoamerica, and they have little resemblance to Peruvian cultivars (Zhang et al. 1998, Zhang et al. 2004). Since varieties of sweet potato were introduced into the Pacific by early European traders and missionaries and by modern agricultural agencies, it remains to be seen how much the DNA studies can tell us.

If a South American Indian raft did bring the sweet potato into the Pacific, it brought nothing else that left a trace. The two most likely explanations for the occurrence of the sweet potato in the Pacific are the deliberate introduction via Polynesian canoes returning from a round trip to the Americas or natural dispersal by drifting capsules or a combination of both. Sailing to America would have been a major challenge for Polynesians, as the voyage is predominantly upwind. Still, it is a feat that Polynesians were capable of, given that return voyages of similar dif-ficulty and magnitude were necessary to reach New Zealand. If the sweet potato was introduced by humans, it is surprising that it was not accompanied by maize, a much more valuable New World staple.

Long-range natural dispersal remains a distinct possibility. The sweet potato seldom flowers when grown in temperate regions, but it does when transported to a tropical climate. The seeds of the sweet potato (*Ipomoea batatas*) are impervious to salt water. Although they sink, they are enclosed in capsules that are buoyant. Closely related species of *Ipomoea* have wide natural distributions across the tropical regions of the Pacific and Indian Oceans. One species occurs on Carpentaria Island in tropical Queensland where the tubers were collected by Aboriginal Australians (Sauer 1993).

CONCLUSION

The peopling of the South Pacific Islands has intrigued scholars since the days of colonial exploration when European sailors found that virtually every island in this vast region was inhabited. Scholars came to realize that the peoples of the Pacific shared morphological traits strongly resembling Southeast Asian peoples. Extensive archaeological studies point to an ancient occupation of Australia and western Melanesia and a much more recent and rapid colonization of some of Melanesia and all of Polynesia.

There are clear archaeological, linguistic, and genealogical bonds between Polynesians of the eastern Pacific and the peoples of Southeast Asia. Polynesian mothers share close genetic links to women in coastal populations throughout the islands of Southeast Asia, in particular those of Indonesia and Papua New Guinea. More distant ancestors of Polynesian women are found throughout the islands and mainland of Southeast Asia. A greater proportion of the paternal ancestors of Polynesians is derived from the adjacent area of Melanesia. The genetic evidence supports evidence of cultural exchange between people in these locations, adding further support to the hypothesis that the Polynesians essentially island-hopped into the Pacific.

It appears that initially the Polynesians paused long enough between settlements for Melanesian males to become increasingly prevalent in the colonizing expeditions. It is likely that there was a tight genetic bottleneck during the early phase of Pacific colonization among both males and females. This is revealed in the small number of male and female DNA lineages among contemporary Polynesians. Most early colonizing voyages probably involved a relatively small number of individuals. As a consequence, the Polynesians are genetically homogeneous and share a close genetic relationship with their relatives to the west. Currently there is scant molecular evidence for migrations from the Americas to Polynesia. The detection of two individuals, out of a sample of 655, who had maternal lineages found in the Americas (Sykes et al. 1995) allows for a remote possibility of intrusion by Native Americans into Polynesia, although it is also true that an Asian origin for these lineages cannot be ruled out.

9

The Outcasts of Israel

And it shall come to pass in that day, that the Lord shall set his hand again the second time to recover the remnant of his people, which shall be left, from Assyria, and from Egypt, and from Pathros, and from Cush, and from Elam, and from Shinar, and from Hamath, and from the islands of the sea. And he shall set up an ensign for the nations, and shall assemble the outcasts of Israel, and gather together the dispersed of Judah from the four corners of the earth.

—Isaiah 11:10-12

With the divine backing of New World scriptures, Latter-day Saints believe that Israelites accomplished at least two marathon oceanic voyages to the New World in approximately 600 BC. Lehi, the head of the principal group, is believed to have been a direct descendent of Israel's son Joseph, who was carried away captive into Egypt. The leader of the other migration, Mulek, is believed to have been the son of Zedekiah, the last king of Judah, who was captured by the Babylonian king Nebuchadnezzar. A number of other Israelite men and women of anonymous tribal affiliation joined both Lehi's and Mulek's migratory parties. By about AD 400, the descendants of these lost Israelites had multiplied into million-strong civilizations and spawned other migratory groups that went on to colonize additional territory in the Americas.

During its infancy, the science of anthropology was frequently preoccupied with the search for lost tribes of some form or other. It is likely that LDS beliefs would have been particularly satisfying to early American scholars and theologians obsessed with the search for the Lost Tribes of Israel. Centuries of misinterpretation and speculation have yielded numerous implausible claims of Hebrew influence in the language, physical appearance, ethnohistory, and customs of native groups in America and Polynesia. But none of these claims has weathered critical examination or convinced dispassionate observers that Israelites beat Columbus to the New World or settled the Pacific before Cook arrived there. As the sciences have shed light on man's biological, cultural, and geographical history, the potency of the lost tribe mythology has waned.

The question of whether or not Jews or members of the Ten Lost Tribes anciently found their way to the New World is susceptible to examination using DNA technology:

> In the same way as religions spread more by conversion than by the sword, genes travel across the globe with the more or less willing participation of those who transmit them. History is a story of love as much as war. Mass movement—of Lost Tribes or anyone else—is not needed ...
>
> Genes can trace a nation's history further into the past than can any record, even one as venerable as the Old Testament. The new map of the world's genes hints at a surprising truth: there may indeed have been a lost tribe, millions strong, emerging from the Middle East to lose its cultural, if not its biological, identity as it moved across the globe. The peoples bound together by this common genetic history range from the Atlantic to the Bay of Bengal and from the North Cape to Ceylon. (Jones 1996)

Scientists have the molecular tools to trace human genealogies long after memories fade and records decay. More importantly, this technology reduces our reliance on the visible similarities between ancient cultures—similarities that can arise purely by chance—and focuses our attention on the central question of whether or not Hebrews reached the western hemisphere before Columbus. Before we can begin to search for Israelites who lost their way while spreading to the four corners of the earth, we need to understand their genealogy and where the Israelite branch is situated in the human family tree.

THE HOUSE OF ISRAEL

While the ancient patriarch Abraham is considered the founding father of the House of Israel, the familiar genealogical divisions among his descendants stem from the male offspring of his grandson Jacob. Renamed Israel by heavenly decree, twelve of Jacob's descendants, including his ten sons, Reuben, Simeon, Judah, Dan, Naphtali, Gad, Asher, Issachar, Zebulun and Benjamin, and two of his grandsons, Ephraim and Manasseh, inherited portions of Palestine and their descendants became grouped into tribes. Dispersals and repatriations of these tribes extend back as far as the founding of the House of Israel.

The Israelites conquered and occupied Canaan during the second millennium BC. Centuries later, they became partitioned into the southern kingdom of Judah, comprising the tribes of Judah and Benjamin, and the northern kingdom of Israel with the remaining tribes. The Babylonians captured the southern kingdom in 586 BC, initiating the dispersal of Judah. When the Persian king Cyrus the Great conquered the Babylonians in 538 BC, he allowed the Jews to return to Palestine; however, most were already scattered among the nations of the Middle East. By the conclusion of the conflict between Palestine and the Romans in AD 70, most of the descendants of Judah resided within the Roman Empire.

The Assyrians captured the ten tribes of the northern Israelite kingdom in 722 BC. Exactly where the Ten Lost Tribes went after their escape from the Assyrians, and their current whereabouts, remains a powerful mystery. According to biblical prophecy, the diaspora must reach the four corners of the earth (Isa. 11:10-12). Believers in the literal truth of this ancient prediction have been anxious to see these tribes united in the land of Israel before the Messiah returns. Since the descendants of the lost tribes and the kingdom of Judah share a common culture, history, and ancestry, we should expect this to be revealed in their genes. To understand the genealogy of this branch of the human family tree, it helps to look further into the prehistory of the Middle East, which has been at the epicenter of the birth of civilization.

The early Caucasoid ancestors of Middle Eastern and European populations began spreading north from Africa about 40,000 years ago (Straus 1989). Their arrival preceded the disappearance of the Neanderthals, an extinct species that shared many of our morphological features (Krings et al. 1997). It is clear that modern humans lived alongside Neanderthals in Europe for at least 10,000 years. The early settlers lived in caves and were hunter-gatherers, a fact most beautifully illustrated in the cave art in southwestern France and Spain (Valladas et al. 1992). The major limitation on human habitation in Europe during this period was the southern extent of glacial fronts during the Ice Age.

About 10,000 years ago, dramatic cultural changes began to occur in the Old World. These changes were accompanied by the spread of agriculture from the Fertile Crescent, the region in the Near East stretching from the Nile and Jordan Valleys through southeast Turkey to the Zagros Mountains in Iran. Local plants (wheat and barley) and animals (sheep, goats, cattle, and pigs) were progressively recruited into growing farming economies. From this region, agriculture steadily spread out in all directions; however, the precise manner in which it spread from the Fertile Crescent is still much debated. One theory contends that the people who developed agriculture slowly radiated out from the Near East, swallowing up or displacing the hunter-gatherer societies that lay before them. At the other extreme, the advance is seen more as the spread of ideas and trade in crops and animals than as a migration of people (Dennell 1983). The truth is likely to lie between these two extremes. It is certain that the early dispersion of humans and ideas from the Middle East had a dramatic effect on the genetic history of mankind.

The latitude and physical geography of Europe and the Mediterranean provided a relatively accommodating environment for human occupancy after the retreat of the ice sheets. Farming arrived in the Middle East 11,000 years ago, in central Europe about 7000 BP, and the base populations from which the current European nations sprang were firmly established by about 4000 BP. Fixed villages and towns appeared in the land of Canaan between about 8500 and 4000 BC, among them Jericho and Megiddo. The first Semitic people on the scene were the Amorites, who penetrated Canaan from the north between 3000 and 2000 BC (*Encyclopedia Britannica*). Other invaders of the region included the Egyptians, Hyksos, and the Philistines, who seem to have originated in Crete. The Israelites entered Canaan in about 1250 BC and settled in the hills to the south. After conquering the Philistines and the native Canaanites under the leadership of King David in 1000 BC, Canaan became known as the Land of Israel, Israelites tracing back both culturally and genetically to the people occupying this small geographic region approximately 3,000 years ago (Bright 1981).

The sacred magnetism of Jerusalem for Jews and Arabs runs deeper than a simple case of disputed geography. The common attachment they feel for the Holy Land and for Jerusalem includes, at its heart, strong biological roots. Blood ties link the nations and ethnic groups living in close proximity to modern Israel. Some of these groups, including Arabs, Samaritans, and the Druze, claim a common ancestor in the patriarch Abraham. The Druze are a distinct cultural group numbering about 250,000 people living mostly in Lebanon and in several communities in Israel and Syria. They have maintained an insular community in the Middle East throughout the last thousand years.

The genealogical ties between Jews and their neighbors reflect a pattern seen around the world. One feature of the emerging map of human genes is that people share the most genetic similarity with their closest neighbors (Cavalli-Sforza et al. 1994). Geography and genes, more often than not, are intimately connected. Unlike the Jews, several Middle Eastern populations have remained in Palestine for the last three thousand years. Knowledge of their genetic makeup helps shed light on the genetic makeup of the Israelites. Next to geography, the most revealing clue to a shared ancestry is a closely related language. Hebrew belongs to the Semitic family of languages which are spoken in Ethiopia in northern Africa, the Arabian Peninsula, the Middle East, and Malta. Other Semitic languages include Arabic, Syriac, and Aramaic. Among the similarities these languages share is the way that consonants are formed at the back of the mouth and throat, indicating a common origin.

Given the Jews' deep spiritual ties to Palestine, it is somewhat ironic that they have spent most of their history in exile. Of the estimated 14 million Jews living today, most are derived from two ethnic groups known as Ashkenazim and Sephardim, distinguished by their most recent place of exile. Ashkenazic Jews, by far the most numerous (~80%), have resided in northeastern Europe for centuries, particularly in the Rhineland. Sephardic Jews number about 700,000 and previously lived around the Mediterranean, predominantly in Spain (Bonné-Tamir et al. 1978). The two communities are culturally linked (Tikochinski et al. 1991) even though they have been in relative isolation from each other during the past 500 years (Bonné-Tamir et al. 1986). Most Sephardic Jews now share present-day Israel with a similar number of Ashkenazim.

The genetic resemblance of Semitic populations adds further weight to the linguistic and geographical support for shared ancestry. Research on a large number of nuclear genes, mostly blood proteins, has led scientists to conclude that Jews are more closely related to other Semitic populations than they are to European people or to the more distant African populations. However, somewhat unexpectedly, Ashkenazic and Sephardic Jews share closer genetic ties with each other than they do with groups in neighboring Semitic communities (Bonné-Tamir et al. 1986, Szeinberg 1979). In spite of their disparate histories, both have maintained a relatively high degree of isolation from surrounding foreign populations.

JEWISH MOLECULAR GENEALOGIES

Paternal and maternal DNA genealogies display the strong genetic links Middle Eastern groups have with Europe. Virtually all of the individuals in Middle Eastern populations have maternal DNA lineages found frequently among Europeans. These European lineages have been classified into distinct lineage families on the basis of specific DNA mutations (Torroni et al. 1994d, 1996), shown in Table 9.1.

Table 9.1 Maternal DNA lineages among Middle Eastern and European populations.

	Maternal Lineage (%)							
Population	H, V	J	K	T	U	X	N	minor
Ashkenazi Jews	**26**	**8**	**32**	**5**	**6**	**1**	**10**	**4**
Near East	31	10	5	10	17	3	2	20
Spain	69	7	5	4	7	1	0	4
United Kingdom	56	16	7	7	8	2	0	3
Europe	**54**	**10**	**6**	**8**	**16**	**2**	**<1**	**4**

Source: Richards et al. 1996, Gonzalez et al. 2003, Behar et al. 2004. The Ashkenazi Jews included 565 individuals from 15 different populations in Europe. The Near Eastern population included individuals from Turkey, the Fertile Crescent from Israel to Iran, the Arabian Peninsula, the lower Nile, and northern Sudan.

The most common female line in Europe is the HV group, which occurs at a frequency of almost 70 percent in Spain. A characteristic of the Ashkenazim populations is the high frequency of the K lineage (32%), typically occurring at low frequencies in most other European populations. The X lineage is rare among Ashkenazi Jews, about 1% (Behar et al. 2004), but reaches its highest frequency, about 25 percent, among a close neighbor, the Israeli Druze (Macaulay et al. 1999).

The X line branches from the N lineage, which is found in about 10 percent of Jews. Among the minor female lineages appearing in the Near and Middle Eastern populations are the L and M groups, both of which usually occur at a frequency of less than 1 or 2 percent. The L lineage occurs in about 76 percent of Africans, and the M lineage is found in about 56 percent of Asians. Jewish populations have clearly been through significant bottlenecks in their history (Behar et al. 2004). However, strong genetic ties to European populations are readily apparent.

Other research on several geographically separated Jewish groups has confirmed the abundance of European maternal lineages in Jewish groups (Thomas et al. 2002). This research also revealed that most Jewish communities were founded by relatively few women and that the subsequent genetic influence of surrounding populations was fairly limited on the female side. This finding is consistent with the tradition that Jewish status, in the absence of conversion, is defined by maternal descent (*Encyclopedia Judaica*).

Just how and when European and Middle Eastern maternal lineages diverged was studied by a team led by Martin Richards from Oxford University (Richards et al. 2000). Most European and Middle Eastern maternal lineages appear to have been derived from a Paleolithic population that expanded into Europe after the worst of the last Ice Age, 20,000 years ago. The ancestors of people from Europe and the Middle East became separated between 10,000 and 20,000 years ago. The spread of Neolithic farmers from the Fertile Crescent beginning about 10,000 years ago probably contributed to about a quarter of the surviving European lineages.

Jewish and non-Jewish males from the Middle East are closely related, sharing similar Y chromosomes. As with the ancestry of their mothers, their paternal lineages are closely related to those of Europeans (Hammer et al. 2000). As one moves from Western Europe toward the Middle East, there is a steady increase in the frequency of the most common Jewish paternal lineages and a corresponding decrease in the frequency of the most common lineage in Western Europe. This strongly suggests a significant population movement between the two regions. Since the trends indicate an east-west orientation, scholars have interpreted this as evidence of the spread of agriculture from the Fertile Crescent sometime after about 10,000 years ago (Rosser et al. 2000). The forebears of Europeans and Semitic groups probably became separated about 20,000 years ago (Richards et al. 2000, Hammer et al. 2000). Significant mixing between the two regions probably peaked during the spread of agriculture and appears to have remained relatively stable since then.

Clearly, Middle Eastern populations represent branches of the European bough of the human family tree. In some cases it is possible to differentiate between Israelite and European lineages and thus distinguish Israelite ancestry from European ancestry. The Y chromosome is particularly suited for this type of research because it is packed with information that can be tapped to identify Israelite-specific DNA lines. A remarkable demonstration of this capacity comes from work among Jews who, through tradition, traced their ancestry back to the ancient patriarch Moses.

SONS OF AARON

According to the biblical record, after the exodus from Egypt in approximately 1500 BC, Moses instigated an important patriarchal tradition among the tribe of Levi when he set apart the male descendants of his brother Aaron to serve as priests (Exod. 28:4). Instead of land, Aaron's descendants inherited special religious responsibilities associated with the House of Israel's portable temple, the tabernacle. It has been Jewish custom since the time of Aaron for priestly responsibilities to be passed down from father to son. Jews inheriting this responsibility are known as the Cohanim or Cohen Jews. Frequently, Cohen Jews have surnames such as Kohn, Cohen, or Kohen that serve as a perpetual reminder of this genetic heritage.

The strict father-to-son inheritance of priestly responsibility mimics the inheritance of the Y chromosome, raising an intriguing question. Is there a unique Y chromosome lineage among Cohens that could have survived the 120 or so generations since Aaron? This question occurred to Karl Skorecki, a member of the Cohanim who is head of molecular medicine at Rambam Hospital in Haifa, Israel. During a moment of reflection at the synagogue, Skorecki was inspired to search for the existence of a clearly distinguishable Cohen Y chromosome. Males who are the direct patrilineal descendants of Aaron, according to Skorecki, might have preserved Aaron's Y chromosome or some closely related chromosome with whatever slight changes may have accumulated during the passage of time.

Based on surveys of Jewish gravestones, about 5 percent of male Jews around the world belong to the priestly tribe (Thomas et al. 1998). Skorecki and his colleagues tested Cohanim, Levite, and Israeli Jews of Ashkenazic and Sephardic origin for a range of unique DNA changes on their Y chromosomes. Remarkably, they found that about 50 percent of Cohens in both ethnic groups possessed virtually identical Y chromosomes. This molecular surname was found in about 15 percent of Israeli Jews and 5 percent of Levites but was essentially absent in non-Jewish Semitic populations. The occurrence of the Cohen DNA lineage in Israelites may point to the movement of Y chromosomes from the Cohanim into the greater population. Another enticing possibility is that the Cohen lineage may be the Y chromosome of the genealogical father of all Israelites, Abraham, who is understood to have lived about 500 years earlier than Aaron. The research shows conclusively that the inception of the Jewish priesthood predated the division of world Jewry into the Ashkenazic and Sephardic ethnic groups over 1,000 years ago.

Skorecki's team found further intriguing evidence that the Cohen Y chromosome may have belonged to Aaron. In addition to the Cohen lineage, several Y chromosomes were identified among Jews that were strikingly similar to the Cohen lineage and almost certainly descended from it (Thomas et al. 1998). The team reasoned that these new lineages probably arose through the occurrence of more recent mutations of the Cohen Y chromosomes. Since the approximate rate of mutation in the Y chromosome is known, it was possible to estimate the time when the original ancestral Cohen Y chromosome existed in a single individual. This was calculated to have been approximately 3,000 years ago, a date that corresponds very well with the biblical account of Moses and Aaron living about 3,300 years ago.

It is now possible to distinguish subgroups of very closely related Cohen lineages derived from a common ancestor who lived within the last few hundred years. Steve Strauss, a personal friend and professor of forestry at Oregon State University, believed that he had Cohen ancestry and took a DNA test to determine if this was the case. The test not only confirmed his possession of the Cohen lineage, it revealed that his Y chromosome was virtually identical to those of two other men who had been recently tested. The three "DNA Cohens" corresponded with each other and found that their paternal ancestors had immigrated to the United States from a small region within thirty miles of the city of Busk in the Ukraine. These three men now know that they are related even though the paper evidence for this may never be found. We are entering a powerful new era of genealogy when relatives may be brought together via the DNA contained in a cheek swab.

AFRICA'S LOST TRIBE

Discovery of the Cohen Y chromosome sparked an investigation of a tantalizing Lost Tribe claim from southern Africa. In Zimbabwe, a black Bantu-speaking people numbering about 50,000 had claimed to be descended from Jews who had traveled to Africa centuries earlier. Known as the Lemba, their oral tradition was of ancestors arriving by boat from a lost city called Sena and that the original party consisted entirely of males who were shipwrecked off the east coast. The Lemba claim to Jewish ancestry was based on scant evidence but included tribal customs such as circumcision, food taboos, and use of biblical names. On the surface, their customs could be Judaic or derived from Muslim or Afghani cultures.

The Lemba said their ancestors built Great Zimbabwe, an abandoned stone-walled city in southern Zimbabwe that has been dated to the thirteenth century. Many early settlers believed that Great Zimbabwe was constructed by a prehistoric white race. However, many scholars considered the ruins to be the work of local indigenous groups (Ndoro 1997).

The controversial claim lured scientists to examine the Lemba more closely. In 1997 Tudor Parfitt, a scholar at the University of London, began a fascinating investigation to see if he could find the lost city of Sena (Parfitt 1997). Traveling from village to village and following clues contained in oral traditions, Parfitt traced a possible route of migration to the east coast of Africa, where he lost track of the presumed trail. Later, while researching the Jews of Yemen, he discovered a city named Sena in the eastern Hadramaut, a valley in southern Yemen. Several local tribes had the same or similar tribal names as those found among the Lemba.

In light of these findings, scholars decided to see if there was in fact a Jewish presence in the paternal genealogies of the Lemba by comparing Lemba, Bantu, and Semitic Y chromosomes (Thomas et al. 2000). It was discovered that a surprisingly high proportion of Lemba Y chromosomes had Semitic origins. About 70 percent of Lemba Y chromosomes are Semitic and the remaining 30 percent are common among surrounding Bantu populations. About one in ten Lemba male lineages proved to be virtually identical to the Cohen paternal lineage—powerful evidence that Lemba oral traditions were based on historical facts rather than myth. The evidence strongly suggested that the Lemba are Jews. However, at this stage it is not possible to rule out a significant contribution from Yemeni Arabs.

Further support for the Lemba oral tradition comes from the way the Lemba society is organized. It is divided into several tribes or clans including the Buba, Hajji, Hamisi, Mhani, Sadiki, and Thouhakale. Each of these clans has a high proportion of individuals with Semitic Y chromosomes. However, most of the Cohen Y chromosomes are found among members of the Buba clan, where it occurs in about half the males (Thomas et al. 2000). According to some Lemba historians, the Buba clan must have come from Judea and migrated to Yemen, where they settled and built the city of Sena. The Buba were thought to have ruled over the other clans. Some believe that the Lemba are the descendants of the sons of Senaah named in the Old Testament (Ezra 2:35), who returned from Babylon in about 537 BC. The city of Sena may have been a halfway home for the paternal ancestors of these black Jews.

ISRAELITE DNA IN THE AMERICAS AND POLYNESIA

If Israelites found their way to America and Polynesia before Columbus, the genetic signs of their presence would have been carried among their living descendants. A survey of the DNA lineages of Middle Eastern populations has revealed that Israelite lineages largely resemble those of Europeans. If members of the House of Israel did enter the Americas, European lineages would hold the most promise for evidence of a connection to the Old World. However, scientists who have detected these rare lineages among native peoples in the Americas and Polynesia assume that they came via sailing ships in the wake of Columbus.

Several lines of evidence support this assumption about the recent arrival of Old World female lineages in the Americas:

1. European female lineages have been found most frequently in North American tribes, particularly tribes that collided most heavily with early colonists. Only a handful have been found south of northern Mexico. In Mesoamerica, considered by some Mormons to be the most likely location of the large Book of Mormon civilizations, native tribes are essentially devoid of European lineages.

2. European lineages occur at very low frequencies in tribes separated by thousands of miles. There are no obvious "hot spots" that would be present if large Hebrew populations were present in the Americas as recently as AD 400.

3. The most abundant European female lineages found among

Native Americans are those that are most common in European countries that played a prominent role in the post-Columbus colonization of the New World.

4. African lineages have been detected in some tribes. These extremely rare lineages are also likely to have resulted from recent admixture, mirroring the pattern of European admixture detected by DNA studies.

About 10 percent of Native American Y chromosomes are likely to have originated in Europe or Africa and seem to have found their way to the New World after Columbus. These lineages are more common in North America and are scattered throughout the two continents, a distribution consistent with a recent origin. Given the wealth of information contained on the Y chromosome, it should soon be possible to determine the particular region of the Old World from which these lineages originated.

European DNA lineages are particularly rare in Polynesia, where it appears that both male and female pedigrees are firmly welded to the Orient. Non-Asian maternal lineages appear to derive from the western end of the European gene pool and are scattered throughout the Pacific at very low frequency. Pacific molecular pedigrees firmly link Polynesians with Southeast Asians, consistent with considerable archaeological, anthropological, and linguistic research that supports a virtually exclusive Asian presence in the ancestry of Pacific Islanders.

The ancestors of Native Americans were Asians who unknowingly became the first Americans as they walked across Beringia over 14,000 years ago. The ancestors of the Polynesians were Asians who honed their considerable nautical skills among the islands of Southeast Asia before sailing out into the Pacific during the last 3,000 years. Regardless of coincidental cultural, linguistic, or morphological parallels with the Old World, the peoples of the Pacific Rim who met Columbus and Cook were not Israelites. They were descendants of a far more ancient branch of the human family tree that had existed thousands of years before the Israelite branch sprouted into existence.

The Troubled Interface between Mormonism and Science

10

The Lord's University

The war between science and religion is best seen as a battle between a shark and a tiger. On its own territory, each is invincible: but stray into the opponent's kingdom and the enemy is bound to prevail.

—Steve Jones, 1996

SCIENCE AND RELIGION

Atop a hill behind St. Peter's Basilica in Rome sits Casina Pio IV, a temple-like structure set amid the splendid gardens of the Vatican. The casina was erected in the 1550s as a retreat for Pope Pius IV but now serves as the home of the Pontifical Academy of Sciences. The academy's membership includes eighty eminent scientists—twenty-five of whom are Nobel laureates—from diverse disciplines, nationalities, and religious beliefs. Formed in 1936 by Pope Pius XI, the purpose of the academy is "to promote the progress of the mathematical, physical and natural sciences and the study of epistemological problems relating thereto" (Seife 2001). Every other year the academy meets to discuss and debate important issues such as environmental concerns, the implications of genetics, and the origins of life and the galaxies. Each time they meet, the pope gains the undivided attention of some of the world's brightest scientific minds. The Catholic church pays for academy meetings, and those who attend are expected to display respect for the work of the church. But speakers choose their own topics and debate issues with complete freedom. The academy has been credited with influencing church policy toward new technologies such as the use of recombinant DNA. The pontiff's decision to make the recent declaration that evolution "is more than just a theory" was almost certainly influenced by the Pontifical Academy (Seife 2001).

Much of the writing on the relationship between science and religion has portrayed these two human activities in militaristic language as polarized, aggressive entities in continual conflict (Brooke 1991, Lindberg and Numbers 1986). Instances of conflict are not hard to find, but it is misleading to say that theologians and scientists have been broadly engaged in protracted warfare. A basic weakness of the thesis is the portrayal of science and religion as discrete entities when, in fact, both involve different expressions of human concern and the same individuals often participate in both (Brooke 1991). At another extreme, apologetic writers have sometimes argued that religion fosters scientific inquiry or that science and religion will reach complete agreement when both are properly understood. Increasingly, historical scholars have concluded that neither "conflict" nor "harmony" appropriately captures the interaction between these two complex social activities (Lindberg and Numbers 1986).

In some quarters there is an acceptance that science and religion should coexist peacefully and respectfully in separate spheres, that the task of science is to define the natural universe, whereas the goal of religion is to produce a moral framework for the world. Stephen Jay Gould, formerly a professor of zoology and geology at Harvard University, viewed the two spheres as "non-overlapping magisteria" or domains of authority in teaching, resurrecting the odd cliché to illustrate his point that "science studies how the heavens go, religion how to go to heaven."

He continued:

> Science tries to document the factual character of the natural world, and to develop theories that coordinate and explain these facts. Religion, on the other hand, operates in the equally important, but utterly different realm of human purposes, meanings, and values—subjects that the factual domain of science might illuminate, but can never resolve. Similarly, while scientists must operate within ethical principles, some specific to their practice, the validity of these principles can never be inferred from the factual discoveries of science. (Gould 1999)

Gould proposed a respectful noninterference, accompanied by intense dialogue, recognizing that the two magisteria cover two very different and critical facets of human existence. The enemy is not science or religion "but dogmatism and intolerance" by the scientists or the religious. The young-earth creationists, according to Gould, exemplify the transgression he identified by attempting to ban the teaching of evolution in public schools and promoting the teaching of a fundamentalist version of life's history. Most scientists and religious leaders have distanced themselves from the fundamentalist movement, in particular from creationists who blunder through the scientific sphere misappropriating evidence to support a ten thousand-year-old earth and a strict creation period of six twenty-four-hour days.

The boundaries between the scientific and religious realms are blurred in Mormon theology since the Latter-day gospel claims to embrace all truth. Science and religion are simply two different instruments for finding, understanding, and disseminating truth, according to Mormon thought. No conflict is believed to exist between true science and true religion. Consequently, if science conflicts with true religion, then the science is incorrect. In his much-cited work, *Mormon Doctrine,* Elder Bruce R. McConkie wrote that "there most certainly is conflict between science and religion, if by science is meant the theoretical guesses and postulates of some organic evolutionists, or if by religion is meant the false creeds and dogmas of the sectarian and pagan worlds" (McConkie 1979).

Not only does Mormonism reserve the right to identify scientific truth, it declares unashamedly that it corners the market on truth in the religious realm. All other religions on the earth are false, according to the belief that has reverberated through Mormonism since Joseph Smith claimed the following conversation with Jesus Christ in 1820: "I was answered that I must join none of them [the Christian churches of the day], for they were all wrong; and the Personage who addressed me said that all their creeds were an abomination in his sight; that those professors were all corrupt; that: 'they draw near to me with their lips, but their hearts are far from me, they teach for doctrines the commandments of men, having a form of godliness, but they deny the power thereof'" (JS-History 1:19).

All other sects in Christendom are believed to be apostate, having broken away from the truth during the Great Apostasy, a term commonly used among Mormons to denote the Dark Ages. We find an equally polarized portrayal of the latter-day religious landscape in a vision delivered by an angel to the Book of Mormon prophet Nephi while en route to the New World in 600 BC. The angel says: "Behold there are save two churches only; the one is the church of the Lamb of God, and the other is the church of the devil; wherefore, whoso belongeth not to the church of the Lamb of God belongeth to that great church, which is the mother of abominations; and she is the whore of all the earth" (1 Ne. 14:10).

Not surprisingly, Mormons are often given the cold shoulder by other Christian denominations, particularly those currently losing converts to Mormonism. The LDS church downplays scriptures such as the one quoted, countering with the more politically correct claim that all churches have portions of the whole truth, which is nevertheless only found completely within the Mormon gospel.

Ironically, while the church claims to possess the full truth, the current generation of Mormons is taught a selective view of church history, omitting what are considered to be the less faith-promoting portions. Speaking to a group of church historians, apostle Boyd K. Packer stressed that "some things that are true are not very useful"

(Packer 1981). Mormonism has more than its fair share of embarrassing skeletons in its closet, but in the Lord's church, history is made to be inspiring.

A rather audacious example of this sanitizing approach to church history appeared in a 1997 church manual entitled *Teachings of Presidents of the Church: Brigham Young*. The manual mentioned Young's first marriage, the date of his first wife's death, and the date of his second legal marriage. The manual not only failed to mention Young's twenty-seven wives and more than fifty offspring, it deliberately avoided the slightest indication that he had lived a polygamous life. Many of the rising generation of Mormons and recent converts to the church remain blissfully unaware of the details of Mormonism's most colorful leader who boldly stood up to presidents of the United States on the issue of plural marriage. Through such censorship of information, the radical nineteenth-century church has faded into twenty-first century respectability.

Even the highly sacrosanct temple ceremonies have been heavily edited to make them less challenging to modern sensitivities. Gone is the non-Mormon minister who once acted as a hireling of Satan to lead people away from the true gospel. Church historians are not expected to notice these changes and are discouraged from looking at previous versions of the temple ceremony. Members are counseled not to speculate about why the changes occurred, and most members dutifully obey. Senior church leaders would prefer that members not ask why changes take place, the result being a thoroughly revisionist history and a fluid quality to doctrine so that it can be modified, added upon, or removed altogether to suit the moment. It would appear that the test of official doctrine is whether it is faith promoting and facilitates conversion and a minimal rate of departure from the church.

THE LORD'S UNIVERSITY

It would be unthinkable for leaders of the LDS church to sponsor non-Mormon scientists in any kind of forum resembling the Pontifical

Academy. If the Brethren were to receive advice from scientists, it would be from Mormon scientists within a Mormon setting. Brigham Young University (BYU), the largest religiously funded university in the United States, is fully owned and controlled by the LDS church. Most of the school's budget comes from tithes contributed by church members, and about 98 percent of the student body and faculty are LDS. The links between the "Lord's University," as it is sometimes called, and the "Lord's Church," are practically seamless. The church influence over the university is omnipresent and omnipotent, the board of trustees consisting of the First Presidency and half of the Quorum of Twelve Apostles. The remaining three positions are filled by the president of the women's Relief Society organization, the president of Young Women, and a male secretary. Twelve of the thirteen-member board sit shoulder to shoulder on the dias during the church's semiannual General Conference.

Unquestioning obedience to ecclesiastical leaders is expected and almost universally adhered to by faithful Mormons, and it is considered unseemly to question church leaders, especially in public. As a result, academic studies at BYU are subordinated to devotion to Mormon leaders and their pronouncements. Most of the faculty and student body would not raise an eyebrow over a preemption of research by school officials. However, like most universities, BYU states that its purpose is to discover, organize, store, and teach the truth in a setting of academic freedom (Bateman 1997). But this occurs under conditions appropriate for the Lord's University, as recently pointed out by the school's president, Merrill Bateman:

> In the search for truth, the university [BYU] has a responsibility to provide an environment in which the discovery of truth is fostered. Two environmental conditions are essential for the discovery process to be as fruitful as possible. The first is freedom of belief or freedom of thought. ... Individual freedom lies at the core of both religious and academic life. Freedom of thought, belief, inquiry, and expression are crucial no less to the sacred than to the secular quest for truth.

> Historically, in fact, freedom of conscience and freedom of intellect form a common root, from which grow both religious and academic freedom.
>
> We believe that faculty members should be able to research and teach in their disciplines without interference as long as a second environmental condition is allowed to operate. ... The second environmental condition concerns the right of an academic institution to pursue its defined mission and to be free from outside control. Every university (public, private, religious, or nonreligious) places limitations on individual freedom in order to create an environment that fosters the discovery and dissemination of truth within the context of its institutional mission. (Bateman 1997)

The mission of BYU is intertwined with the mission of the church, and individual freedom comes with strict limitations. Public behavior that contradicts or opposes rather than analyzes or discusses fundamental church doctrine is not tolerated either by the university or by the leaders of the church (Bateman 1997). Criticism directed toward the leadership is particularly objectionable. No opinion is tolerated that would embarrass the Brethren. In addition, the faculty and students must receive an annual endorsement from off-campus church officials in order to be involved with the university. Anyone who fails to receive this roadworthy certificate finds himself or herself out the door and on the road to another university and their contract or enrollment at BYU terminated.

The friction between faith and freedom reached a breaking point during the late 1980s and 1990s when several faculty were fired, the circumstances of their dismissals having received press attention. In 1997 the American Association of University Professors (AAUP) considered a number of these cases while specifically investigating the firing of Gail Houston, an English professor who was refused tenure for advocating prayer to a mother in heaven (AAUP 1998). Mormon theology includes belief in the existence of a heavenly mother but "unofficially" discourages prayer to her, although it is presumably now "officially" inappropriate. As university president Bateman explained, "We do not believe that they should be able to publicly endorse positions contrary to the doctrine or to attack the doctrine." This leaves faculty in the position of having to second-guess Mormon leaders, who are increasingly loath to make unequivocal statements of doctrine. The ultimate responsibility to determine the doctrine—and the damage, always with the benefit of hindsight—remains firmly in the hands of the General Authorities who sit on the university's board of trustees.

While the AAUP expressed concern about the minefield of church doctrine, a more disturbing finding surfaced during interviews with a large number of faculty and some of the graduate students. Many, who typically requested anonymity, described instances of censorship at the university, which led the investigative committee to conclude that the "climate for academic freedom [at BYU] is distressingly poor" (AAUP 1998). The AAUP chose to censure the university. This constituted a minor slap on the wrist but was nevertheless a setback for a university that is concerned about national credibility and respectability. Similarly, during BYU's 1998 re-accreditation review, evaluators saw indications of poor staff morale and concern about academic freedom.

Students, like faculty, are expected to abide by strict standards of behavior and even to dress and groom themselves in a manner "consistent with the dignity adherent to representing The Church of Jesus Christ of Latter-day Saints." The Student Honor Code, which governs the required behavior of BYU students on campus and in their private lives, leaves little room for maneuver. The earring quota for women is one per ear. For male students, it is none. Piercing of other parts of the body is out of the question. As expected, strict rules govern behavior in the student accommodations on campus. Off-campus, failure to maintain the BYU Residential Living Standards similarly subjects one to disciplinary action. The Honor Code office encourages students to report those who violate standards, and apparently many students oblige. The most persuasive motivation is undoubtedly the annual endorsement provided by a student's own bishop. In any case, the emphasis is on conformity rather than on personal freedom.

LDS INDOCTRINATION

In contrast to the trends observed in other institutions of higher learning, the BYU student body became more conservative during the second half of the twentieth century. Much of the credit for this goes to the Church Educational System (CES) which expanded dramatically during this period. The CES operates a unique worldwide "seminary" program of one hour of instruction, five days a week, for students in the last four years of high school. An "institute" program provides similarly styled instruction to college-aged students. The church holds fast to the proverb, "Train up a child in the way he should go: and when he is old, he will not depart from it" (Prov. 22:6). Globally, seminary students study the same lesson material and spend a year each on the Old and New Testaments, LDS church history, and the Book of Mormon. Some students attend seminary during school hours, but outside Utah, most attend before school. Classes often begin as early as 6:30 a.m., placing considerable strain on conscientious students in the most critical years of high school. Parents who express concern about this have often found the CES personnel to be inflexible, insisting that gospel learning takes precedence over secular learning.

Church officials carefully vet CES curricular material, and seminary and institute teachers are strongly counseled to avoid intellectualizing or engaging students in controversial issues. Armand Mauss, an LDS sociology professor at Washington State University, observed that "historical and contemporary studies of the Church Educational System would almost certainly demonstrate in great detail the gradual (and probably deliberate) transition from a pedagogical philosophy of intellectual articulation and reconciliation to one of indoctrination" (Ostling and Ostling 1999). The CES has become increasingly parochial, anti-intellectual, anti-scientific, and intolerant of those who question its policies and approach.

The course material for CES deliberately avoids difficult issues that apologists deal with on a regular basis. The current university-level CES Book of Mormon institute manual steers students away from any attempt to link historical details in the book with factual research into the Americas. In one example, students are provided with a mud map of positional relationships between sites mentioned in the text but are counseled that "no effort should be made to identify points on this map with any existing geographical locations." The manual quotes extensively from Book of Mormon defender Hugh Nibley, but the considerable body of work of LDS anthropologist John Sorenson is entirely overlooked. Theories of a minimal Hebrew impact in the New World advanced by the Foundation for Ancient Research and Mormon Studies (FARMS) are much too controversial for inquiring young minds. Students indoctrinated with the belief that most of the ancestors of Native Americans are Lamanites find nothing to confront this assumption in the student manual. Church president Spencer W. Kimball's words remind students that back in 1977 there were "some half million Indian or Lamanite members in the Church." In another quote, Kimball makes reference to the millions of Lamanites who farm the steep hillsides of the Andean ranges and serve in menial labor in Ecuador, Chile, and Bolivia, as well as the deprived and untrained Lamanites who live in North America.

For Mormons who choose to deepen their faith in the Book of Mormon through scholarship, learning about the limited archaeological evidence to support the book can be a shock. Because of years of exposure to popular Mormon mythology in seminary and Sunday School classes, most LDS youth enter the university having never considered the Book of Mormon to be anything other than a literal, hemispheric account of the ancestors of essentially all Native Americans and Polynesians. Rather, they have learned that archaeological discoveries in the New World support the Book of Mormon. They may have heard the name of Thor Heyerdahl, but such respected Gentile scientists as Brian Fagan, Francis Jennings, Peter Bellwood, and Luca Cavalli-Sforza will be unfamiliar to them. The concept of Clovis hunters roaming a continent several thousand years before the date assigned to Adam and Eve in the Garden of Eden is difficult for them to accept.

MORMONISM AND SCIENCE

Few Mormons ever come face to face with research that threatens their traditional beliefs. The closest contact for almost all members is through second-hand accounts from church leaders or LDS apologists who tell them how scientists continue to get it wrong. There is a strong taboo in Mormonism that prevents most members from reading books that are suspected of being critical of the church or challenging to its doctrines. This, combined with the fact that most members follow their leaders without question, ensures that members are effectively immunized against anything that undermines their fundamentalist beliefs. Gentile scientists who discover uncomfortable truths are commonly suspected of being in cahoots with "anti-Mormons."

This distrust of scientists' motives is most palpable with regard to those who uncover evidence challenging perceived religious truths. Biologists have been the most frequent trespassers upon territory that Mormons believe has been illuminated by divine revelation. This branch of science has been brutal with traditional Mormon beliefs linked to man's early history on earth. Mormon theology, given further authoritative clout in the form of revelations to Joseph Smith, supports a literal interpretation of most biblical events such as the creation of man, the Tower of Babel, and the Flood.

The church struggles with evolutionary theory where the origin of human beings is not given wide enough berth. The perception among Mormons is that the church has officially denounced evolution. This is not surprising given the vitriolic "unofficial" attacks senior Mormon leaders have directed its way. Bruce R. McConkie, of the First Council of Seventy, subjected the theory to a ten-page dressing down in his book *Mormon Doctrine,* long considered an authoritative source for Latter-day Saints, especially after McConkie became a member of the Quorum of the Twelve Apostles. McConkie concluded that "there is no harmony between the truths of revealed religion and the theories of organic evolution" (McConkie 1979).

Senior apostle Boyd K. Packer has been equally scathing, hurling salvos at the theory from the safety of a General Conference pulpit rather than allowing open dialogue on the subject. He has said that "no idea has been more destructive of happiness, no philosophy has produced more sorrow, more heartache and mischief, no idea has done more to destroy the family than the idea that we are not the offspring of God, only advanced animals, compelled to yield to every carnal urge" (Packer 1992).

Mormon apologists point out that such statements are not doctrine and that the church has not officially denounced the theory. But most Mormons find it difficult to disregard proclamations by prophets and apostles from Salt Lake City, especially statements made at the pulpit during General Conference.

At the same time, BYU biologists face overwhelming evidence almost daily in their research, so that they have generally come to accept an evolutionary explanation for the diversity of life on earth. Nevertheless, it is rare that ecclesiastical leaders acknowledge these views from informed BYU scholars; when they do, it receives little publicity (Jeffery 1973). Biologists and those in related fields resolve the issues to their own personal satisfaction. In contrast to General Authorities, the statements made by LDS scientists in support of evolution are expressed timidly and within the context of faith. The most widely known contemporary LDS defense of evolution was expressed by a non-biologist who was convinced by the mounting evidence his colleagues had accumulated. In *Reflections of a Scientist,* Henry Eyring, who is the father of current apostle Henry B. Eyring, voiced his contentment with a common ancestry with apes as long as he knew that God had been at the controls:

> God has left messages all over in the physical world that scientists have learned to read. These messages are quite clear, well-understood, and accepted in science. That is, the theories that the earth is about four and one-half billion years old and that life evolved over the last billion years or so are as well established scientifically as many theories ever are.

> We should keep in mind that scientists are as diligent and truthful as anyone else. Organic evolution is the honest result of capable people trying to explain the evidence to the best of their ability. From my limited study of the subject I would say that the physical evidence supporting the theory is considerable from a scientific viewpoint.
>
> In my opinion it would be a very sad mistake if a parent or teacher were to belittle scientists as being wicked charlatans or else fools having been duped by half-baked ideas that gloss over inconsistencies. That isn't an accurate assessment of the situation, and our children or students will be able to see that when they begin their scientific studies. (Eyring 1998)

Notice that Eyring wisely avoided singling out senior leaders of the church for the bad press that evolution has received in LDS circles. In spite of Eyring's efforts, Mormons generally dismiss the theory and hold to the biblical chronology, believing the human race originated with the recent arrival of Adam and Eve.

Latter-day Saint scholars feel less compelled to drag science to the depths occupied by the Creation Science movement (Nelkin 1982). This small group of scientists from diverse specializations has sought out scientific information that either supports the literal biblical chronology or appears to threaten the theory of organic evolution. They staunchly believe in a recent, sudden creation of the universe and a literal interpretation of all major events in the Old Testament (Strahler 1987). While many Mormons consider the earth to be a few thousand years old, most educated Mormons have not felt the need to reduce the earth's age to a few biblical millennia. A central tenet of Mormonism is the belief that there are "worlds without number" (Moses 1:33) and that the universe is infinite in size and age. It has always existed and always will, according to this concept. Consequently, a few billion years for creation, which Mormons consider to be more of an organizational process, is not difficult to swallow. A common position occupied by Mormons is that the earth is ancient and underwent a long evolutionary period, followed by a "special creation" introducing mankind into a world ready to be subdued.

Another notable piece of insight concerning man's genesis is the revelation Joseph Smith received regarding the Garden of Eden, that instead of having been in the Old World, it was in fact in North America in Jackson County, Missouri. This land is considered sacred ground because it is "the land where Adam dwelt" (D&C 116; 117:8, 11). In the last days, a new Zion, the "New Jerusalem,"will be built there, and the Saints will gather in Jackson County before Christ's return. Mormons have had little difficulty accepting that man's early ancestors migrated enormous distances across the globe. But in this case, theology has made things a little easier. Travel between the New World and the Old World was feasible before the Flood because there was only a single land mass on the earth, according to Mormon teachings (D&C 133: 23-24). The continents are believed to have come into existence after the Flood.

Mormons are compelled to believe in a literal Flood that covered the entire earth symbolizing the earth's baptism, something that in Mormonism requires complete immersion in water. If such a catastrophe had occurred, it would have caused one of the greatest extinctions of plant and animal life in the earth's history. Science has not been kind to this myth. No evidence has been found to confirm that such an event occurred during the last few thousand years of the earth's history. The difficulty for Mormons is accentuated by the fact that, in addition to the Bible, modern scripture dictated by Joseph Smith (Moses 7:34-51) added further support and clarification to the biblical story. Through the loins of Noah sprang "all the kingdoms of the earth," according to the revelation. Members are assured that a universal flood, followed by a division of the continents, is not inconsistent with a faithful and correct interpretation of scientific evidence. Recently the official *Ensign* magazine claimed that misguided interpretation of geological evidence has led some scientists to reject the historicity of a universal Flood:

> Both of these groups—those who totally deny the historicity of Noah and the Flood and those who accept parts of the story—are persuaded in their disbelief by the way they interpret modern science.

> There is a third group of people—those who accept the literal message of the Bible regarding Noah, the ark, and the Deluge. Latter-day Saints belong to this group. In spite of the world's arguments against the historicity of the Flood, and despite the supposed lack of geologic evidence, we Latter-day Saints believe that Noah was an actual man, a prophet of God, who preached repentance and raised a voice of warning, built an ark, and floated safely away as waters covered the entire earth. We are assured that these events actually occurred by the multiple testimonies of God's prophets. (Parry 1998)

Parry was privately criticized by scientists at BYU who were disturbed by the poor standard of the scientific arguments portrayed in this article. Many of these scholars probably accept that Noah was a real man but that the Flood was a localized event, elaborated upon by ancient scribes. There is reliable evidence that a flood of mammoth proportions inundated the Black Sea about 7,000 years ago (Ryan et al. 1997). However, the source of the deluge was in fact the Mediterranean Sea rather than the heavens. Rising sea levels after the last Ice Age are thought to have caused a breach in the land between the Mediterranean and the Black Sea, resulting in a massive inundation of seawater through what is now the Bosporus Strait.

Mormons who are becoming familiar with molecular genealogy will be inevitably exposed to similarly "controversial" Gentile research concerning the origins of the human family. The growing DNA data fit within a vast repertoire of research that is revealing an increasingly coherent picture of our past. Mitochondrial, Y, and recent X chromosomal research is reinforcing the already intensively documented view of our ancestors originating from a population that lived in Africa roughly 100,000 years ago. The substantial scientific evidence for such an ancient, African origin of all human beings is virtually ignored by Mormons. Most are not sufficiently informed about, or lack the belief structure to cope with, widespread settlement of the Americas in 11,000 BC. Latter-day Saints will find it difficult to reconcile this with the biblical Tower of Babel and the Flood, nor will they know where Adam and Eve can be grafted onto such an ancient human family tree.

BOOK OF MORMON APOLOGETICS

The front line of defense for the church has been occupied since 1979 by scholars associated with FARMS, originally a privately funded organization, but predominantly staffed by Mormons employed at BYU. In recognition of the close ties with the university and their capacity to "provide strong support and defense of the Church on a professional basis," President Gordon B. Hinckley officially guided FARMS onto BYU campus in 1997. It now exists under the umbrella of another group called the Institute for the Study and Preservation of Ancient Religious Texts.

Through FARMS, the church is now able to mobilize a considerable number of scholars to respond to critics of church positions, typically branded as "anti-Mormons." Book of Mormon apologetics have dominated the published work of FARMS, although the organization has attempted to broaden its range of issues. All FARMS research is published on the assumption that the historicity of the Book of Mormon need not be questioned. It is an assumed fact that the Book of Mormon contains an authentic history of the ancestors of pre-Columbian American Indians. FARMS has proven to be highly successful in spurring an increase in the volume of scholarly investigations into the Book of Mormon in the form of books, journals, articles, newsletters, updates, and reviews of books. On volume alone, the critics have fallen far behind. This has led senior researcher Noel Reynolds to claim that FARMS now sets the agenda (Reynolds 1999).

One of the most popular outlets for rebutting Book of Mormon critics has been the *FARMS Review*, a clearinghouse for anything that broaches the subject of the Book of Mormon. Reviews of General Authorities' books and those of the FARMS staff are predictably favorable. On the other hand, when an unsympathetic work is examined, the FARMS microscope switches to high magnification to be sure that readers apprehend every blemish. The intention is obviously to deter members from reading any book that challenges their faith. These are not regular book reviews, but substantial works of scholarship that are copiously referenced. For example, in 1993 Signature Books, a perennial thorn in the side of FARMS, published a ten-essay compilation edited by Brent Metcalfe that questioned the Book of Mormon's historicity from a number of perspectives (Metcalfe 1993). The response was an entire issue of the *Review* answering what FARMS considered to have been a multi-pronged attack. The apologist counter claims ran over 500 pages, much longer than the offending tome itself. Mormon scholarship supporting the Book of Mormon appears in the *Journal of Book of Mormon Studies*, a FARMS-BYU publication. Such scholarship rarely reaches externally reviewed journals. Consequently, FARMS remains an insular group that preaches to the choir.

11

Plausible Geography

The first rule of historical criticism in dealing with the Book of Mormon or any other ancient text is, never oversimplify. For all its simple and straightforward narrative style, this history is packed as few others are with a staggering wealth of detail that completely escapes the casual reader. The whole Book of Mormon is a condensation, and a masterly one; it will take years simply to unravel the thousands of cunning inferences and implications that are wound around its most matter-of-fact statements. Only laziness and vanity lead the student to the early conviction that he has the final answers on what the Book of Mormon contains.

—Hugh Nibley, 1952

In August 1921 James E. Talmage, a scientist and member of the Quorum of the Twelve Apostles, received a letter from a non-Mormon scholar inquiring about the Book of Mormon. The scholar, a Mr. Couch from Washington, D.C., was perplexed by certain aspects of the book that appeared to be in conflict with what was known about ancient America. Talmage passed the troubling letter to one of the First Council of Seventy, Brigham H. Roberts, who was widely respected for his analytical study of the Book of Mormon and well deserving of his unofficial title of "Defender of the Faith." Roberts was charged with the responsibility of finding suitable answers to the questions raised in the letter.

Until this time, the Book of Mormon had largely escaped serious comparison with secular theories about the origins of "Indians" and the archaeological discoveries in the Americas. At the time of its first appearance, the Book of Mormon was not attacked on the basis of contradiction with the prevailing views because, in fact, it reflected the popular and even much of the learned thinking about America's original founders. Many of the early nineteenth-century scholars had speculated about an Israelite origin for the American Indians. As scientific study increased during succeeding decades and the popular mythology was exposed, Mormons were largely preoccupied with the hardships encountered while migrating across the Great Plains and establishing settlements in the arid West. A protracted fight with the United States government over polygamy soon followed, alienating Mormons from the United States. The Rocky Mountain Saints also found themselves relatively isolated from scientific research institutions and centers of learning, which at that time were almost exclusively in the eastern states.

Mr. Couch was troubled by the issue of how the language spoken by Book of Mormon people in the fifth century AD could have so rapidly multiplied into the staggering diversity of languages observed among Native Americans one thousand years later. He was also perplexed by the mention of horses, steel, "cimeters" (Persian sabers from the 16th-18th centuries AD) and silk—all undetected in the New World societies that had greeted the Europeans after 1492 (Smith 1984). Roberts was persuaded enough by the questions to find it absolutely necessary to address these problems, "as it is a matter that will concern the faith of the youth of the Church now as also in the future" (Roberts 1992).

Within a few months, Roberts had completed a three-part manuscript examining the difficulties raised by Mr. Couch and other issues Roberts uncovered during his studies. Over three days in early 1922, he stoically presented these to church president Heber J. Grant and members of the Quorum of the Twelve and First Council of Seventy. Roberts was hopeful that a discussion of these troubling issues with the enlightened members of the Quorum of the Twelve—and with much-anticipated divine inspiration—would produce solutions for the problems. He was disappointed by the results of the conference which provided little, if anything, that assisted in the defense of the Book of Mormon. No answers were given, and the Twelve dutifully stood one after the other and bore testimony to the truthfulness of the book.

Further study of Book of Mormon difficulties by Roberts included a comparison of the book's narrative with the popular beliefs in the area where Joseph Smith grew up and an examination of some of the book's internal inconsistencies. Ethan Smith's 1825 book, *View of the Hebrews: Or the Tribes of Israel in America,* was particularly troubling for Roberts, who saw extensive parallels between it and the Book of Mormon. Roberts produced two manuscripts: "Book of Mormon Difficulties" (1921) and "A Book of Mormon Study" (1923). Both were clearly intended for publication but did not reach the printers until 1985 when members of the Roberts family donated copies of the manuscripts to the University of Utah and Brigham Young University (Roberts 1992). The latter manuscript contained a grave assessment of the Book of Mormon, in which Roberts concluded that a nineteenth-century origin was entirely plausible:

> It will appear in what is to follow that such "common knowledge" did exist in New England; that Joseph Smith was in contact with it; that one book, at least, with which he was most likely acquainted, could well have furnished structural outlines for the Book of Mormon; and that Joseph Smith was possessed of such creative imaginative powers as would make it quite within the lines of possibility that the Book of Mormon could have been produced in that way. (Roberts 1992)

Some among the Mormon intelligencia have maintained, presumably untroubled by such comments, that Roberts was a faithful believer who was playing devil's advocate (Madsen and Welch 1985).

During the remainder of the twentieth century, Mormons rallied to bring scholarship to bear on the Book of Mormon just as Roberts had hoped. Mormon scholars have, in fact, found a "staggering wealth of detail" in the book which, as Hugh Nibley observed, "completely escapes the casual reader." This effort has been aimed largely at reinterpreting the book to align the narrative with a modern understanding of New World colonization. The upshot has been a steady contraction of LDS claims regarding the scale and geographical footprint of the Israelite presence.

GLOBAL GEOGRAPHY

New clarifications of the Book of Mormon story are largely preoccupied with the question of geography. Where were the Jaredite and Lehite civilizations located on the American continent? Fortunately, among the most straightforward portions of the narrative are its descriptions of the relative positions of people jostling with each other for room in the Promised Land. The landmark with which all Book of Mormon civilizations can be positioned is a particularly conspicuous geographical feature mentioned frequently in the text. All of the major settlements can be positioned in relation to a "narrow neck of land" flanked by an eastern and western sea and connecting two land masses.

> And thus the Nephites were nearly surrounded by the Lamanites; nevertheless the Nephites had taken possession of all the northern parts of the land bordering on the wilderness, at the head of the river Sidon, from the east to the west, round about on the wilderness side; on the north, even until they came to the land which they called Bountiful.
>
> And it bordered upon the land which they called Desolation, it being so far northward that it came into the land which had been peopled [by Jaredites] and had been destroyed, of whose bones we have spoken, which was discovered by the people of Zarahemla, it being the place of their first landing.
>
> And they came from there up into the south wilderness. Thus the land on the northward was called Desolation, and the land on the

southward was called Bountiful, it being the wilderness which is filled with all manner of wild animals of every kind, a part of which had come from the land northward for food.

And now, it was only the distance of a day and a half's journey for a Nephite, on the line Bountiful and the land Desolation, from the east to the west sea; and thus the land of Nephi and the land of Zarahemla were nearly surrounded by water, there being a small neck of land between the land northward and the land southward.

And it came to pass that the Nephites had inhabited the land Bountiful, even from the east unto the west sea, and thus the Nephites in their wisdom, with their guards and their armies, had hemmed in the Lamanites on the south, that thereby they should have no more possession on the north, that they might not overrun the land northward. (Alma 22:28-32)

This description brings to mind an hourglass-shaped landmass, something that was firmly conceptualized in the minds of the Lehite writers by about 90 BC (Alma 50:34; 52:9; 63:5) and upon which the entire Book of Mormon narrative is overlaid. Similar descriptions are found in the account of the Jaredites (Ether 10:20), which comes later in the book but earlier in time, their arrival from the Middle East occurring in about 2000 BC. The Nephites refer to the northernmost lands as Desolation because they are covered by the decaying ruins of the earlier Jaredite civilization. The Nephites inhabit a land immediately south of Desolation known as Bountiful, which is just north of Zarahemla—the latter being the land that was also settled by the Mulekites. The narrow neck of land falls between Bountiful and Desolation. Throughout most of the Book of Mormon, the Lamanites occupy the southernmost land of Nephi.

Mormons have consistently seized upon the obvious similarity between this hourglass landscape and the hemispheric geography of the two American landmasses. Consequently, most believe that the "land northward" and the "land southward" correspond to the continents of North and South America and that the narrow neck of land is the Isthmus of Panama at the northern tip of South America. The isthmus is thirty miles across at its narrowest point, a brisk "day and a half's journey for a Nephite." All church presidents and other General Authorities have essentially subscribed to this view, as do most of the current members. This is clearly how apostle James E. Talmage pictured the geography of the book in his classic work, *Jesus the Christ*—one of few books published by the church itself—as he elaborated about the expansion of the Lamanites after the final destruction of the Nephites in AD 400:

> They spread northward, occupying the northern part of South America; then, crossing the Isthmus, they extended their domain over the southern, central and eastern portions of what is now the United States of America. The Lamanites, while increasing in numbers, fell under the curse of darkness; they became dark in skin and benighted in spirit, forgot the God of their fathers, lived a wild nomadic life, and degenerated into the fallen state in which the American Indians—their lineal descendants—were found by those who rediscovered the western continent in later times. (Talmage 1915)

The LDS church continues to teach that Native Americans are the direct descendants of Book of Mormon people. The introduction to Book of Mormon editions since 1981 states explicitly that the Lamanites are "the principal ancestors of the American Indians." Since the traditional geography model most closely aligns with this statement and with an uncontrived reading of the Book of Mormon, it is not surprising that it is still the most widely accepted view in the church.

Despite wide acceptance by leaders and members of this global view of Book of Mormon geography, most "serious" Book of Mormon scholars, particularly those at Brigham Young University, maintain that this hemispheric geography is out of the question. The scholars at BYU have experienced great difficulty in trying to align descriptions of travel times, population growth, and the geographical proximity of events with the vast territories of North and South America. Throughout the 1,000-year history of the Nephites and Lamanites, their major population centers were relatively fixed within several days march of each other. One would expect cultures of the type described in the

Book of Mormon to have left significant traces of their presence. For example, they were remarkably advanced in writing, agricultural practices, and widely used metallurgy. They possessed wheel technology, practiced Christianity, built Christian temples, and established cities with populations of tens of thousands. In contrast, the land adjacent to the Isthmus of Panama lacks the remains of any large, complex civilization.

LIMITED GEOGRAPHY

Dozens of alternative models of geography have sprung up over the years (Clark 1992). However, there is only one serious contender accepted by most Mormon academics, which proposes that most Book of Mormon events took place in a restricted part of Mesoamerica. Only in Mesoamerica are there ruins of civilizations of the magnitude evident in the Book of Mormon.

LDS scholars support this local or "limited geography" approach to Book of Mormon topography as presented by Professor John L. Sorenson of Brigham Young University. Sorenson argues that the only acceptable narrow neck of land that meets the requirements of the Book of Mormon is the Isthmus of Tehuantepec in southern Mexico (Sorenson 1985). What is more, he asserts that all serious LDS students of the Book of Mormon who have grappled with the problem in recent decades have reached this same conclusion. The Lehite lands, according to his view, must have been restricted to a 400-mile-long section of Mesoamerica that spans the cultural region of southern Mexico and northern Central America (Figure 11.1). There are obvious difficulties with the Isthmus of Tehuantepec, not the least of which is that a 125-mile crossing, as the crow flies, is a formidable "day and a half's journey" on foot. Another glitch is that the east and west seas mentioned in scripture have to be shifted almost 90 degrees because they are essentially south and north of the narrow neck of land.

The limited geography model creates other problems for scholars such as the need to have a prominent hill for the Jaredite extermination and the final battles of the Nephites and Lamanites. The hill, which the Nephites called Cumorah and the Jaredites called Ramah, was near the narrow neck of land and major population centers. The hill where Joseph Smith obtained the gold plates containing the Book of Mormon is also called the Hill Cumorah, and most Mormons believe that the climactic battles of both civilizations took place there in upstate New York, thousands of miles to the northeast of Mesoamerica. Disciples of a hemispheric geography justifiably assume that the Hill Cumorah in the Book of Mormon and the Hill Cumorah in New York State are one and the same hill. The Cumorah of the Book of Mormon contained buried records and was located "in a land of many waters, rivers and fountains" (Mormon 6:4), a landscape remarkably similar to upper New York state, a region teeming with lakes.

Academics have had to make sense of these contradictions and ex-

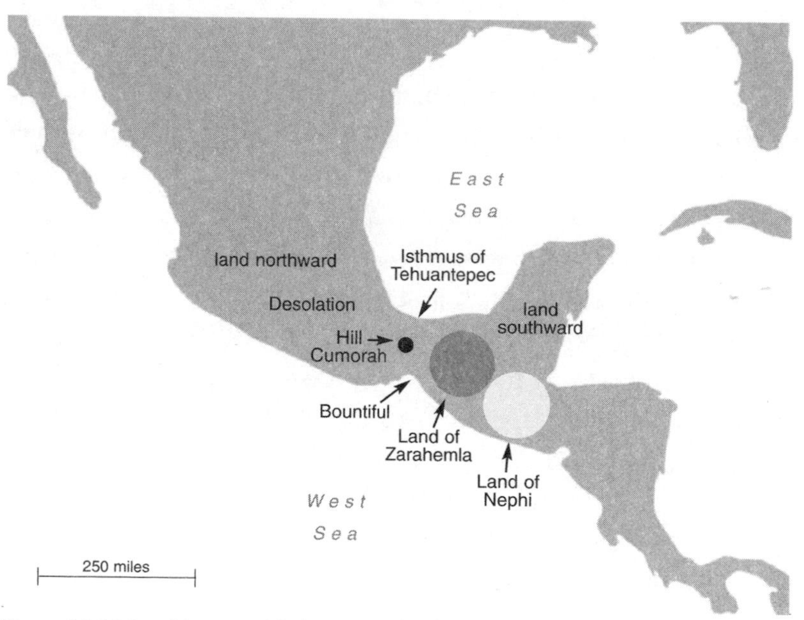

Figure 11.1 Most Mormon scholars consider the geographical territory covered by the Book of Mormon to be limited to Mesoamerica. According to this view, the "narrow neck of land" is located at the Isthmus of Tehuantepec in southern Mexico. The "land northward" and "land southward" are thought to correspond roughly with the Mesoamerican cultural area.

pose the "cunning inferences" that escape casual readers. Why would hundreds of thousands of Lamanites and Nephites march from Mesoamerica to New York to fight a final battle of extermination? To account for this anomaly, Mormon scholars have concluded that there are in fact two Cumorahs. The Hill Cumorah referred to in the Book of Mormon is not the one in New York State from which the gold plates were recovered. The scriptural Cumorah is, in fact, in Mesoamerica where most of the Nephite records were deposited and where the final Nephite slaughter took place (Sorenson 1985). The task of transporting the gold plates the four thousand miles from the Mesoamerican Cumorah to the hill in New York fell to Moroni, according to the revisionists, as the lonely last surviving Nephite prophet. It has been suggested that Moroni traveled by canoe from Central America to New York via the Gulf of Mexico, the Mississippi River, and the Ohio River during the last years of his life (Sorenson 1985). There is no mention of this incredible transcontinental trek in the Book of Mormon, the closing chapters of which are said to have been recorded by Moroni.

Another impetus for the restricted geography is the obvious fact that the Americas were widely inhabited thousands of years before the arrival of the Jaredites in 2200 BC. The astounding array of cultures and languages encountered by early Europeans could not have originated from the Hebrew said to have been spoken by the Nephites and Lamanites of AD 400. The only plausible explanation for Mormon apologists is that the two groups of Semitic immigrants—the Lehites (Lamanites and Nephites) and Mulekites—occupied a restricted area in the Americas.

OTHER PEOPLE IN THE NEW WORLD

An important consequence of this compression of the geography and acknowledgment of the presence of non-Book of Mormon peoples is having to explain how the large numbers of native peoples who lived throughout the Americas interacted with those described in the golden-plate account. Unfortunately, the Book of Mormon offers little assistance in this regard. There is no indication in the record that the Jaredite or Lehite parties came into contact with any native people whose origin could not be accounted for in the book:

> If such other races or tribes existed then the Book of Mormon is silent about them. Neither the people of Mulek nor the people of Lehi or after they were combined, nor any of their descendants ever came in contact with any such people, so far as any Book of Mormon account of it is concerned. (Roberts 1992)

The failure of Book of Mormon prophets to mention other races or peoples in the New World was a logjam in Mormon apologetics that was begging to be cleared. Bravely, the irrepressible Sorenson stepped into the breach with his groundbreaking research paper, "When Lehi's Party Arrived in the Land, Did They Find Others There?" (Sorenson 1992). The premise was that major concerns related to the Nephite and Lamanite history could be resolved if references to "others" could be found in the Book of Mormon. The reader was provided with an exhaustive thirty-four-page dissection of the Book of Mormon text in search of "all possible evidences" for "other" people living in close proximity to the Lamanites and Nephites. Not surprisingly, the research achieved its objective, leading Sorenson to conclude that "readers will not be justified in saying that the record fails to mention others."

This style of scholarship is not uncommon among the work of Mormon apologists. It is an understandable practice if one appreciates that the apologists essentially never question the historicity of the Book of Mormon. For them, it is a fact that the Book of Mormon is true history of real people. This research paper has become the most commonly cited work to placate Mormons who are troubled by the lack of any unambiguous reference to the millions of Native Americans who were present in the New World when the Lehite and Jaredite civilizations flourished. For many Mormon scholars, the uncomfortable issue is resolved and can be put aside.

In addition to reinforcing the limited geography theory, the occurrence of "other" people was required to help explain the rapid explosion of Nephite and Lamanite numbers to the point where they were involved in "wars and contentions" within twenty-five years of their arrival in the New World. Sorenson suggests that it could account for the Nephites' preference for multiple wives and concubines (Jacob 2:23-27) and for their construction of an impressive temple (2 Ne. 5:16) soon after arriving in the New World. Sorenson finds the necessary references, a subset of the thousands of inferences Nibley anticipated, that had escaped the notice of readers of the Book of Mormon for decades. Included in his analysis is a rather novel interpretation of a prophecy of Lehi who, after stepping off the boat from Israel, foreshadowed the events that would befall his children:

> [I]t is wisdom that this land should be kept as yet from the knowledge of other nations; for behold, many nations would overrun the land, that there would be no place for an inheritance.
>
> Wherefore, I, Lehi, have obtained a promise, that inasmuch as those whom the Lord God shall bring out of the land of Jerusalem shall keep his commandments, they shall prosper upon the face of this land; and they shall be kept from all other nations, that they may possess this land unto themselves. ...
>
> But behold, when the time cometh that they shall dwindle in unbelief, after they have received so great blessings from the hand of the Lord ... I say, if the day shall come that they will reject the Holy One of Israel, the true Messiah, their Redeemer and their God, behold, the judgments of him that is just shall rest upon them.
>
> Yea, he will bring other nations unto them, and he will give unto them power, and he will take away from them the lands of their possessions, and he will cause them to be scattered and smitten. (2 Ne. 1:8-11)

Mormons generally think that the nations referred to in this scripture are the Gentile Christian nations of Europe, after Columbus, by whom the "Indians" were certainly dispossessed, "scattered and smitten." Sorenson points out that this view requires little imagination, as the scripture does not indicate how distant those nations were and when they would arrive (Sorenson 1992). He proposes that these "other" nations could have been already present when Nephi's boat beached and used immediately as instruments of the Lord against the straying covenant people. The Book of Mormon is predictably silent about any of these non-Jewish peoples, instead giving the impression that Lehi arrived in a land uncluttered by man—a land truly kept "from the knowledge of other nations." The only people guilty of scattering and smiting the Nephites during their 1,000-year history are the Lamanites, who do so regularly.

Sorenson argues that there was plenty of room for "other" people to be absorbed into a lower class, subordinate to the powerful Nephite elite. Similarly, the absorption of "others" could account for the "exceedingly numerous" Lamanites. Scriptural accounts characterizing the Lamanites as wild, naked, blood-thirsty hunters who ate raw meat and wandered in the wilderness (Enos 1:20) are at odds with the fact that the Lamanite population continually grew faster than the more advanced Nephite agriculturalists. The only plausible explanation is that the Lamanites included or dominated other people who lived by cultivation. This base population could supply sufficient food to sustain the large Lamanite civilization and provide "an almost inexhaustible supply of sword fodder" (Sorenson 1985, 1992).

A potential source population for the hypothesized "others" would be survivors of the Jaredites, who had arrived in 2200 BC. However, according to the Book of Mormon, the Jaredite nation was completely destroyed in an enormous battle and left only one lone survivor, Coriantumr. Still, LDS scholars now consider it unlikely that every Jaredite showed up for the senseless final battle (Sorenson 1992, Nibley 1988); besides, the term "destroyed" does not necessarily imply that everyone was killed. Remnant populations may have survived, which shared with the Nephites and Lamanites such cultural practices as the growing of maize (Sorenson 1992). This overlooks the earlier Book of Mormon prophecy where the nature of the destruction is clearly spelled out. Coriantumr is warned by the prophet Ether that if he does not repent (and he did not), his kingdom would be completely destroyed. Ether emphasizes that "every soul shall be destroyed save it were Coriantumr" and "he should only live to see ... another people [Lehites] receiving the land for their inheritance" (Ether 13:20-21).

Sorenson proffers that references to "other" people are vague because these people were outside the focus of the Book of Mormon and because the Nephite record keepers thought them too insignificant to mention (Sorenson 1992). The Native Americans who were present in every corner of the New World well before 600 BC are thereby reduced to an insignificant peasant underclass, recruited to build the Nephite temples and fight in the Nephite-Lamanite wars. As they were not Lehi's descendants, they were beneath mention in a book devoted to a favored branch of the House of Israel. Sorenson wonders if acknowledgement of the millions of people who occupied the surrounding lands would have been seen by Nephite chroniclers as a waste of space on their precious plates. Those familiar with the Book of Mormon narrative may find this last argument especially difficult to accept. At times, the account is particularly repetitious and tedious, as Mark Twain observed when he labeled the book "chloroform in print."

In any case, a radically new interpretation of Book of Mormon history has emerged among Mormon scholars. Many of them have questioned the widely held belief that present-day American Indians are Lamanites. Most LDS apologists now accept that the Americas were widely and heavily populated at the time the Lehites arrived on the continent. FARMS writers propose that while Lehi and his small group quickly dominated the native populations soon after they arrived in the New World, their own populations may never have become numerically significant (Johnson 2003). The shift from a macro-history of all ancient peoples of the continents of North and South America to a micro-history of a few people who lived somewhere in Mesoamerica corresponds with the exponential growth in secular research revealing an overwhelming connection to Asia. For all the criticism leveled at mainstream dogma, the thinking of Mormon scholars is now more aligned with their Gentile colleagues than the teachings of latter-day prophets.

Theories of a limited Lehite impact on the Americas may satisfy some Mormons, but they stand at odds with the relatively straightforward reading of the Book of Mormon that the vast majority of Mormons actually believe. It is hard to imagine that an impartial reader, uneducated in either secular New World research or Book of Mormon scholarship, could escape the perception that the book is about the first Americans, Israelite sailors who reached an uninhabited Promised Land and proceeded to establish great civilizations based on Old World cultures. This literal view requires the least academic exertion and sits most comfortably with the "matter-of-fact statements" in the narrative and statements by generations of LDS General Authorities. Theories about limited colonization contradict the statements of God to Joseph Smith recorded in the Doctrine and Covenants, wherein the "Indians" who retreated before the advancing American colonists are said to have been Lamanites (D&C 54:8). Numerous books written on the subject by LDS scholars reassure Mormons that evidence has been found for the existence of Christian civilizations in ancient America. It is understandable why most Mormons believe that the American Indians descended primarily from Jews and that the Book of Mormon is geographically hemispheric.

The limited success so far in swaying popular LDS opinion is a constant source of frustration for Mormon apologists. Sorenson, the most energetic of the revisionists, has spent decades swimming against the mainstream of Mormon dogma:

> A very common view is that the entire Western Hemisphere was involved and that Nephites and Lamanites were everywhere until the final destruction of the Nephites in New York, and then only Lamanites were left and those were the Indians. Now, it does little for our satisfaction in understanding the scripture to take that simpleminded view. It simply is not true. ... If that were so, if the whole hemisphere were the scene, then the Book of Mormon text, its own statements about itself, and its scene would be wrong. ... (Sorenson 1995)

It appears that Mormons are generally content to picture the Book of Mormon story in a setting that is factually wrong. For most Mormons, the limited geographical models create more problems than they solve. They run counter to the dominant literal interpretation of the text and contradict popular folklore as well as the clear pronouncements of all church presidents since the time of Joseph Smith. Mormon apologists remain a tiny minority in the church, and their views do not carry the weight of the prophetic declarations of the Brethren. For as long as Mormons rely more on feelings and literal traditions than on modern scholarship and reason, these problems will afflict the apologists, who are fighting against the very force that convinces most Mormons that the book is without a doubt true.

12

Faith-Promoting Science

We who have a different heritage should be filling in the gaps, linking up real past and real present—concretely, believably, and truthfully—and not just continuing to construct stories and pageants that we then label "Lamanite." If Latter-day Saints believe the Book of Mormon is real, as they say, they should treat its setting as reality.

—John Sorenson, 1985

Mormon scholars have not been afraid to step into the realm of science for evidence in support of the Book of Mormon or to garner ammunition against what they see as Gentile dogma. Consequently, there is a long history of tension between Mormons and non-Mormon scientists about the colonization of the Americas. This tension has been more pronounced among Mormons due to their greater interest in New World prehistory than Gentile scientists have had in Mormonism. Scholars on the Mormon side have maintained for years that mainstream theories about Amerindian origins are seriously flawed, and even now, criticisms of "hardheaded" Gentile views frequently surface in Mormon writings. In nearly complete ignorance of this persistent theological resistance, Gentile science continues to map out the trail of America's nomadic founders back to their Siberian homeland. While Mormons hold to the doctrine about Native Americans being predominantly—or even partially—descended from a lost Israelite race in the face of overwhelming evidence to the contrary, this strikes outsiders as being willfully stubborn.

BOOK OF MORMON ARCHAEOLOGY

The geographical territory that captures the imagination of LDS intellectuals is Mesoamerica, a land where New World civilization reached its zenith. The early Saints were tantalized by descriptions of the breathtaking Mayan civilizations in John Lloyd Stephens's 1841 bestseller, *Incidents of Travel in Central America: Chiapas and Yucatan*. Joseph Smith found the book so inspiring, he declared Palenque a Nephite city. Modern scholarship indicates that this Mayan center was built after AD 600, over 200 years after the Lamanites exterminated the Nephites; but dating details aside, Mormon scholars continue to find the remains of Mayan cities to be prime candidates for where Lehi's people might have lived. The Maya developed complex cultures, built impressive cities, and were the only literate New World civilization—three critical prerequisites for any candidate for the lands and people described in the Book of Mormon. In addition, the Maya built pyramids that superficially resembled the Old World pyramids of Egypt.

The Jaredite inhabitants of the Americas, from the time of the Tower of Babel until their destruction in 200 BC, according to the Book of Mormon, are usually identified with the Olmec, a civilization that preceded the Mayas. Michael Coe, an authority on the Olmec and one of few Gentile archaeologists to have seriously examined the Book of Mormon, has found no evidence to support this thesis. Coe observed that Mesoamerican archaeology has benefitted tremendously from serious Mormon interest, but the history of Book of Mormon archaeology is littered with apostasy as numerous gifted scholars have been swayed by the views of their Gentile colleagues (Coe 1973).

In the late 1930s, Fawn Brodie, niece of church president David O. McKay, began a short essay on the likely early nineteenth-century sources of the Book of Mormon, little aware of the impact her essay would have on Mormon studies. Over the next few years, her essay evolved into the critically acclaimed biography of Joseph Smith, *No Man Knows My History*, first published in 1945. Her book incensed a bright young Mormon scholar named Hugh Nibley who had recently completed a Ph.D. in ancient history at the University of California (Reynolds 1999). Nibley railed against Brodie's biography in a series of polemical attacks published the same year under the title *No Ma'am, That's Not History: A Brief Review of Mrs. Brodie's Reluctant Vindication of a Prophet She Seeks to Expose*.

For over half a century since then, Nibley has been Mormonism's chief "Defender of the Faith" and has been regarded in some circles as something of a lay prophet. His intellectual tack with the Book of Mormon has been to search for parallels between it and Old World antiquities. His studies have been serialized in church magazines and collected in numerous volumes. Together with the work of John Sorenson, Nibley's prolific writings have reached a wide Mormon audience and sustained the belief that real people, living in a real place in the New World, were the authors of the Book of Mormon. Still, because Nibley's expertise lies outside of Mesoamerican archaeology, his work, at times, has not been well received by other BYU scholars, many of whom have had serious misgivings about his *modus operandi*. John Sorenson, while commenting on the unprofessional writing that is passed off as archaeology in Mormon circles, severely criticized Nibley and Milton R. Hunter, two scholars who frequently trespassed into unfamiliar New World territory:

> Two of the most prolific are Professor Hugh Nibley and Milton R. Hunter; however, they are not qualified to handle the archaeological materials their work often involves ... As long as Mormons generally are willing to be fooled by (and pay for) the uninformed, uncritical drivel about archaeology and the scriptures which predominates, the

> few LDS experts are reluctant even to be identified with the topic. (Sorenson 1966)

In 1986 Kent Jackson, a respected BYU history professor, was asked to review the first volume of the collected works of Hugh Nibley (Jackson 1988). The historian observed that Nibley's tendency was to "presuppose a common worldview for all the disparate cultures of the ancient world" and then pick and choose the evidence that suited his argument. "This is what inevitably happens when scholars let their predetermined conclusions set the agenda for the evidence." Jackson expressed concern over Nibley's use of the "straw man" approach in his scholarship, "frequently misrepresenting his opponents' views to the point that they are ludicrous, after which he has ample cause to criticize them." Such public condemnation of a respected LDS scholar would be almost unthinkable today. Senior LDS academics, particularly those who stridently defend the church, are now afforded much the same kind of respect granted to the Brethren and are rarely criticized in public. Nibley's wit, enthusiasm, and charisma safeguarded the faith of many Latter-day Saints through a difficult period and inspired a generation of apologists.

Among the better-known early Mormon archaeologists was Thomas Ferguson, a man who devoted his life to locating the ruins of cities described in the Book of Mormon. His efforts led to the formation in 1952 of the New World Archaeological Foundation (Larson 1996). Ferguson successfully petitioned the church for considerable financial support for Mesoamerican archaeological digs. He launched his first expedition in January 1953, accompanied by two graduate students from BYU among others (Sorenson 2002).

Ferguson was a lawyer by profession and his approach in *One Fold and One Shepherd* was to argue a legal defense for the Book of Mormon, going "so far as to present his case as a series of legal exhibits that only the most prejudiced and ignorant judge and jury could fail to find convincing" (Coe 1973). But in the end, the evidence failed to convince even Ferguson. After repeatedly failing to locate archaeological remains that could be linked to the Book of Mormon, Ferguson experienced a private crisis of faith (Larson 1996). On the other hand, John Sorenson, one of the two BYU students who had traveled with Ferguson to Mexico, says he has yet to encounter an argument that is troubling to his faith. Sorenson's interest in these issues began while he was serving as a missionary in the Pacific in 1947. He was in the Cook Islands when Thor Heyerdahl sailed on his *Kon-Tiki* raft from Peru to the Tuamotus, and Sorenson was enthralled by the voyage. Since that time, Sorenson has remained a firm believer in the widespread diffusion of humans and their culture between the Old and New Worlds (Sorenson 2002).

A colleague of Ferguson became a General Authority in 1945. Elder Milton R. Hunter was an amateur archaeologist and contributed to the genre with his 1956 *Archaeology and the Book of Mormon* and, in 1970, *Great Civilizations and the Book of Mormon*. Hunter's views spilled over into his sermons at General Conference, where he liked to publicize the myth of the feathered serpent, or Quetzalcoatl, observed throughout Mesoamerica as a representation of Jesus Christ. Leading Mesoamerican archaeologists disagree with this interpretation. Jacques Soustelle notes that the Quetzalcoatl legend reaches back to the Olmec era as early as 1200 BC (Soustelle 1984) and Michael Coe has found an early connection between Quetzalcoatl and the principal Olmec deity, the jaguar god—a deity associated with rain (Coe et al. 1986).

Among Hunter's interests was an artifact he found particularly significant—a stone relief at Izapa near Chiapas, Mexico, now widely known by Mormons as Stele 5. The stele depicts a complex scene of richly dressed figures seated and standing on either side of a tree, a scene depicted elsewhere in Mesoamerica. Wells Jakeman from BYU thought the stone depicted the "Tree of Life" dream that Lehi experienced in the Book of Mormon (1 Ne. 8), a representation of the trials and temptations the faithful must overcome in order to earn an eternal reward. The dozens of parallels Jakeman saw between the monument and the dream have been less than convincing to later generations of LDS scholars (Larson 1996). But Hunter placed great faith in this and in the heavily Christianized accounts of pre-Columbian history recalled by native writers several decades after Pizarro's conquests of Mexico. Most scholars generally qualify such accounts because they appear to have been molded to please the early Christian missionaries.

DIMINISHING CLAIMS

If an early nineteenth-century mind created the Book of Mormon, one would expect cultural and physical anachronisms to emerge in the book because Joseph Smith would have known few of the historical details of pre-Columbian America. We return here to the issues that spurred B. H. Roberts into action in the 1920s and that troubled doubting Thomas Ferguson several decades later. Anachronisms there are, and they continue to haunt Book of Mormon apologists as they busily scour the New World research conducted by Gentiles. This effort has provided barely plausible explanations for most of the difficulties.

According to the Book of Mormon, steel swords and iron metallurgy were common in the Jaredite and Lehite periods. However, evidence of iron or steel smelting in pre-Columbian America has not been established (Coe et al. 1986). Metallurgy involving smelting and casting of gold and silver first appeared in Mesoamerica in about AD 800. The wheel, another technology familiar to Book of Mormon peoples, was never used in ancient America. A wheeled child's toy has been advanced by some apologists as evidence of the existence of the wheel. Even if there were wheeled vehicles, there were no draft animals in Mesoamerica to pull them. There are no archaeological remains of Old World staples such as wheat or barley. Native Americans in Arizona domesticated a barley species. However, it was a species native only to North America and found only in a limited region around Arizona, a long way from the only New World civilizations that could fit the requirements of the Book of Mormon. Some apologetic explanations elicit feelings of pathos, where it has been suggested that references to horses in the Book of Mormon may have been to deer or to the small, pig-like tapir, neither of which was ridden by Native Americans.

Raymond Matheny, a former BYU professor of anthropology, concluded that the scientific evidence does not support the Book of Mormon. After working in Mesoamerican archaeology for twenty-two years, he concluded that the existence of people and events chronicled in the Book of Mormon, whether in Central America or anywhere in the western hemisphere, cannot be supported by scientific evidence (Matheny 1984). His thoughts echo those of Dee Green, also involved in archaeological research at BYU:

> The first myth we need to eliminate is that Book of Mormon archaeology exists. Titles on books full of archaeological half-truths, dilettanti on the peripheries of American archaeology calling themselves Book of Mormon archaeologists regardless of their education, and a Department of Archaeology at BYU devoted to the production of Book of Mormon archaeologists do not insure that Book of Mormon archaeology really exists. If one is to study Book of Mormon archaeology, then one must have a corpus of data with which to deal. We do not. The Book of Mormon is really there so one can have Book of Mormon studies, and archaeology is really there so one can study archaeology, but the two are not wed. At least they are not wed in reality since no Book of Mormon location is known with reference to modern topography. Biblical archaeology can be studied because we do know where Jerusalem and Jericho were and are, but we do not know where Zarahemla and Bountiful (nor any other location for that matter) were or are. It would seem then that a concentration on geography should be the first order of business, but we have already seen that twenty years of such an approach has left us empty-handed. (Green 1973)

Michael Coe, now retired from Yale University, is well acquainted with the history of Book of Mormon research and was equally frank in his assessment. The "bare facts of the matter are that nothing, absolutely nothing, has ever shown up in any New World excavation which would suggest to a dispassionate observer that the Book of Mormon, as claimed by Joseph Smith, is a historical document relating to the history of early migrants to our hemisphere" (Coe 1973). Twenty years later he was of like mind, considering the book to be "a fanciful creation by an unusually gifted individual living in upstate New York in the early nineteenth century" (Ostling and Ostling 1999).

The focus of Book of Mormon apologetics in recent years has been increasingly on internal evidences for the truth of the Book of Mormon since the external scientific evidence has not been forthcoming. Many of these works mimic the style of Nibley, relying on accumulating parallels discovered between the Book of Mormon and various Old World writings and traditions. Since it is assumed that Joseph Smith could not possibly have known about these parallels in 1830, the only reasonable conclusion is that the book is literally true history. Within a boldly titled article, "Mounting Evidence for the Book of Mormon," in a recent issue of the official LDS *Ensign* magazine, Daniel Peterson produced a copiously referenced review of much of this literature, paying particular attention to the internal proofs of the text's antiquity (Peterson 2000). Lower Mexico and Guatemala are cited as plausible locations for the Book of Mormon story.

Over the years the bold claims for archaeological proof of the Book of Mormon have diminished in favor of this kind of search for parallels in the Old World rather than in the New World. Contemporary Mormon scholars have endeavored to pull back from the enthusiastic excesses of earlier generations. Sorenson dismantles two of the most widely cited archaeological evidences used in support of the Book of Mormon—the Izapa stele and the legend of the bearded white god Quetzalcoatl—in a paper pointing out the two-edged nature of some evidences. Sorenson acknowledges that these earlier "proofs" were based on highly subjective interpretations of Mesoamerican myths and artifacts. Quetzalcoatl's less divine attributes were simply overlooked because it would not have been faith promoting to link them with Jesus Christ (Sorenson 1995).

Most of the current leading Mormon archaeologists have acknowledged the dearth of evidence that can be identified as having anything to do with Nephite or Jaredite cultures (Clark 1992, Nibley 1964, 1988, Warren 1990, Johnson 1992). In the *Encyclopedia of Mormonism,* we find a revealing statement that "of the numerous proposed external [Book of Mormon] geographies, none has been positively and unambiguously confirmed by archaeology ... [and should] at best, be considered only intellectual conjectures" (Clark 1992). The evidence "is indirect and cannot be linked to a single person, place or thing mentioned in the text" (Johnson 1992). The implication is that no New World archaeological findings can be solidly connected with the people, events, or locations described in the Book of Mormon. After decades of apologetic archaeology, spawning dozens of books and research papers, "Book of Mormon archaeology" has yielded little, if any, credible evidence.

The continual contracting of the Lehite influence in the Americas is not all bad news for LDS scholars and for the church, as noted recently by apostle Dallin H. Oaks, who reasoned that such a retreat will prohibit opponents of the Book of Mormon's historicity from proving it false. Oaks is a former Utah Supreme Court justice and well aware of the futility of trying to prove a negative. According to Oaks, critics must now prove that Book of Mormon peoples did not live anywhere in the Americas (Oaks 1993). It is, of course, impossible to prove conclusively that Jews never migrated to ancient America, just as it is impossible to prove that giants never lived on the moon. The most that can be said from the evidence is that there is no credible evidence that ancient Jews ever migrated to the Americas.

Despite this quiescent recognition of the paucity of archaeological support for the Book of Mormon, the LDS church does not appreciate attention being drawn to this fact, even by non-Mormons but particularly not by Gentiles at scientific institutions. The Smithsonian Institution in Washington, D.C., has been among the most offensive to the church in this regard. Since 1951 the Smithsonian has produced a handout (see Appendix C) that it sends to enquirers—generally those who have heard a rumor that the Smithsonian uses the Book of Mormon as a guide to scientific explorations in the Americas. The handout lists several of the Book of Mormon's claims that are not consistent with scientific knowledge regarding ancient America. Among the latter are the Mongoloid physical type of the American Indians, the lack of any Old World domesticated food plants and animals, the absence of iron and steel metallurgy, and the absence of any Old World writings in the New World.

For years this statement had irritated Mormons. Eventually it prompted John Sorenson to write a detailed critique of it in 1982 to pacify those who were troubled by its contents. In 1995 he wrote a revised version of this critique, including a recommendation that the "Smithsonian Institution completely modify their statement to bring it up-to-date scientifically." The church then exerted various forms of persuasion to sway the Smithsonian Institution. Members of FARMS met with a representative of the institution, and Mormon members of Congress cited the "inappropriateness of a government agency taking a stand regarding a religious book." In 1998 the institution dropped the statement, replacing it with the following brief note:

> Your recent inquiry concerning the Smithsonian Institution's alleged use of the Book of Mormon as a scientific guide has been received in the Smithsonian's Department of Anthropology.
>
> The Book of Mormon is a religious document and not a scientific guide. The Smithsonian Institution has never used it in archaeological research and any information that you may have received to the contrary is incorrect. (Smithsonian Institution 1998)

Of course, the Smithsonian is entitled to comment on religious claims when they fall within the scientific realm, but it recognizes the importance of its Congressional funding, which in recent years has been threatened. The Smithsonian still stands by the former, more detailed statement, but its Public Affairs department chose to "simplify" it. Never slow to bathe the Book of Mormon in a favorable scientific light, Mormon apologists interpreted the institution's retreat as an indication that the Book of Mormon is no longer thought to be completely inconsistent with Mesoamerican archaeology (Reynolds 1999).

On the subject of Polynesian origins and migrations, the Mormon church also finds itself unhappily exposed by the "learning of men." On balance, the cumulative evidence provides little to suggest a cultural link between Polynesia and the Americas. The archaeological, linguistic, ethnographic, botanical, and recent molecular evidence reveal a coherent picture of Polynesian migration from the islands of Southeast Asia through Melanesia and out into the Pacific. Even Sorenson conceded in 1985 that the belief that "Hagoth reached Polynesia must rely on faith rather than on reliable evidence" (Sorenson 1985). Research during the two decades since then appears to have confirmed his insight. The strongest evidence of a cultural link across the Pacific is the presence of the South American sweet potato in Polynesia. Claims of a linguistic link between "kumara," the South American Quechua word for sweet potato, and its Polynesian names have been disputed. It is probable that the South American name was introduced by European colonizers and traders, perhaps via the Spanish "Manila Galleons" that regularly crossed the Pacific between Mexico and the Philippines from the mid-1500s until the early nineteenth century. It remains to be seen if further research can reveal the precise American source of the puzzling tuber.

13

LDS Molecular Apologetics

Good science does not consist of someone dreaming up a pet theory and then quilting together pieces of evidence to support it from as many disparate sources as possible while conveniently ignoring pieces of evidence that may undercut the theory.

—Michael Whiting, 2003

When Brigham Young entered the Salt Lake Valley in 1847, it was with a party of "140 free men, three women, two children, and three colored servants," according to the Brigham Young Monument on Main Street near the Salt Lake Temple. By the time the temple was completed about fifty years later, the valley's population had multiplied to about a quarter of a million, mostly due to the frequent arrival of wagon trains from the East bringing people from Britain and Scandinavia. Some of these immigrants now have as many as twelve thousand living descendants, and many of them, from the beginning, kept well-documented records. Today these records are among the largest extended family pedigrees in the world.

MOLECULAR GENEALOGY IN UTAH

In the early 1970s, George Cartwright, then chairman of the University of Utah's Department of Medicine, recognized the unparalleled genetic resource that these large family records represented. He began linking the genealogies of Mormons with the medical registers of the state of Utah. Now known as the Utah Population Database, it contains information on about 1.5 million descendants of a founding group of about ten thousand Mormon pioneers (Jones 1996). This invaluable resource has made possible the identification of several disease-causing genes in humans by allowing scientists to link the inheritance of particular traits and genes through multiple generations.

The human genome project relied heavily on this extensive genealogical database. It was at the University of Utah in the early 1980s that Raymond White and Mark Leppert, professors of human genetics, built the first human genetic marker map using DNA information from Utah families. Their map consisted of the ordered arrangement of "sign post" DNA sequences that allowed researchers to identify the location of the genes they were studying. In turn, this map acted as the framework for organizing the sequence of the 3 billion bases of DNA that the human genome project revealed. Consequently, the family histories of Mormons made a major contribution to how we understand human genetics.

The power of molecular biology to answer questions about an individual's genealogy has also captured the attention of LDS scientists. In March 2000, BYU commenced an impressive global molecular genealogy project aimed at welding traditional family histories with cutting-edge DNA technology. Backed by Mormon philanthropists James Sorenson (no direct relation to John Sorenson) and Ira Fulton, the project planned to locate the ancestral homeland of a person who had no recorded pedigree and to do so based on the genetic information stored in the person's blood. At the close of 2003, approximately 40,000 individuals had given blood and genealogical records to the project. Most Mormon volunteers have extensive family pedigrees of deceased ancestors and have had these ancestors posthumously (vicariously) baptized into the Mormon faith.

Perhaps to distance the church from potential negative fallout from the project, ties with BYU and the church were severed and the project relocated off campus to the Sorenson Molecular Genealogy Foundation in Salt Lake City in early 2004. Scientific director of the project, Scott Woodward, has remained optimistic that the work will yield the world's most comprehensive DNA genealogical reference database and allow people to trace their ancestral homelands. The project is inevitably poised to uncover considerable data that could be used to investigate genealogical ties between Native Americans, Polynesians, and Israelites, but James Sorenson has maintained that this would be a minor portion of the project and that if no ties were found, so be it. "We're searching for the truth," he said in 2000, "let the chips fall where they may" (Egan 2000a). FARMS has downplayed the potential of the project. John Sorenson's read on the situation is that DNA evidence may never provide a definitive answer to the question of Indian origins, that maternal DNA studies cannot trace Lehi in any case. "With [DNA] sampling," said Sorenson, "you may or may not find evidence of a connection to the Old World. If you do, that says something. If you don't, that says that more research needs to be done" (Egan 2000b). Given that researchers at the Molecular Genealogy Foundation have determined about 5,000 Y chromosome DNA lineages, it is likely that they can recognize the Y chromosomes of several well known polygamists in church history, including Brigham Young and Joseph Smith. This project is well positioned to reveal surprises in families descended from respectable pioneer stock, which may prove too disconcerting for some.

FIRST IMPRESSIONS

In 2000 FARMS published its initial verdict on the use of DNA to trace genealogical and evolutionary relationships in global populations in two articles published in the *Journal of Book of Mormon Studies*. John Sorenson wrote both articles, although this was not disclosed by the journal. In the first, under the title "Genetics Indicates that Polynesians Were Connected to Ancient America" (Sorenson 2000a), readers encountered a robust sketch of Rebecca Cann's 1994 study (Lum et al. 1994), from which Sorenson deduced that the "genetic ties linking the two areas are now hard enough to support a picture of substantial historical connections between Polynesian and American groups." Sorenson's information came largely from a popularized account of Cann's research published in the *Hawaii Magazine,* and Cann's more recently published work, which had revealed a virtually exclusive genetic bond between Polynesia and the Far East (Lum and Cann 1998, Lum et al. 1998), was not discussed by Sorenson.

In the next issue of the journal, the BYU professor's earlier enthusiasm had evaporated. In "The Problematic Role of DNA Testing in Unraveling Human History" (Sorenson 2000b), Sorenson provided a scathing assessment of the "new toy in human biology and anthropology" for determining issues of human history. He predicted that the usefulness of DNA research would run a "life cycle" from seeming to "sharply modify the conventional picture" to being relegated to its more rightful place somewhere in the grab bag of scientific tools. This opinion served to deflect attention from the fact that DNA research had confirmed, rather than sharply modified, the conventional theories concerning the origin of the human family and Native American and Polynesian origins. The molecular research had only threatened conventional Mormon views on these topics.

Sorenson took aim at "enthusiasts without adequate critical acumen" in an unsavory attack on Cann, whom he had just praised for her work on Polynesians. This time she had struck a sensitive doctrinal nerve. In her ground-breaking study of the mitochondrial DNA of 147 women from four continents, Cann had concluded that women can claim a common mother from sub-Saharan Africa who lived about 200,000 years ago (Cann et al. 1987). A weakness in Cann's work, which other scientists pointed out and Sorenson repeated, was that it focused on a small (400 base pairs of DNA) portion of the mitochondrial DNA that mutates at too high a rate to allow a definitive estimate of age. Recent, more comprehensive, work on both female and male lineages has confirmed the sub-Saharan origin but predicted that our common ancestors lived there 90,000 to 150,000 years ago (see chapter 5). More significantly, this research sits comfortably with a substantial amount of archaeological and anthropological research that paints a similar picture. Sorenson appears reluctant to accept such antiquity for the human race because of its menacing implications for other areas of Mormon theology. Many church members still believe in an Adam and Eve who lived on the earth barely 6,000 years ago, that the flood of Noah's time eliminated most of the human population 4,400 years ago, and that America was first occupied by Middle Easterners 4,200 years ago. Those who hold such views have difficulty with the idea of ancestors who migrated out of Africa over 60,000 years ago and who have lived uninterrupted on all major continents for the last 15,000 years.

Sorenson provided what was for most Mormon readers their first glimpse of female DNA lineage research on Native Americans (Sorenson 2000b). In his disparaging survey, he cited one research paper where four founding female DNA lineages were described, others where seven and then nine were found, and yet another with thirty "distinct lineages." He concluded that scientists now "choose to simplify the confusion by talking about four Amerindian haplogroups—A, B, C, D" and by simply "dump[ing]" the baffling anomalies into an "other" category. Given such assumed sloppiness on the part of researchers, it was easy for Mormon readers to dismiss it.

Then in 2002, Thomas Murphy, a cultural anthropologist specializing in Mormon studies, publicly raised the issue of the historicity of the Book of Mormon based on modern genetics research. In his essay, *Lamanite Genesis, Genealogy, and Genetics,* included in the volume, *American Apocrypha: Essays on the Book of Mormon,* Murphy considered the implications of New World DNA for views on Native Americans (Murphy 2002). He caught the attention of church authorities when he implied that the Book of Mormon was created from the imagination of Joseph Smith:

> From a scientific perspective, the Book of Mormon's origin is best situated in early nineteenth-century America, not ancient America. There were no Lamanites prior to ca. 1828, and dark skin is not a physical trait of God's malediction. Native Americans do not need to accept Christianity or the Book of Mormon to know their own history. The Book of Mormon emerged from Joseph Smith's own struggles with his God. Mormons need to look inward for spiritual validation and cease efforts to remake Native Americans in their own image. (Murphy 2002)

In his doctoral thesis, Murphy paid particular attention to the racial basis of Mormon attitudes about Native Americans. He found that the Indian Student Placement Program, which placed Native American children in urban white Mormon families, was a systematic effort to turn them "white and delightsome."

I should mention my own participation in this discussion about DNA genealogies in ancient America and the Pacific. I wrote an article in 2000 entitled "DNA Genealogies of American Indians and the Book of Mormon," published on the *Recovery from Mormonism* website (Southerton 2000). The article was a non-scholarly, somewhat candid personal account detailing when I encountered the Indian origins problem and how I had reacted to it. It contained a brief summary of the molecular research to that point. In 2002, I presented a more detailed summary of the research on Native Americans and Polynesians in an address at the Exmormon Foundation Annual Conference in Salt Lake City (Southerton 2002). In February 2003, Thomas Murphy and I published a letter in *Anthropology News* calling for a renunciation of the Mormon folklore about Native Americans and Polynesians sharing an Israelite ancestry and carrying a curse because of misdeeds on the part of their ancestors (Murphy and Southerton 2003).

The considerable attention that was devoted to Lamanite DNA during this period resulted in the distribution in 2002 of a video documentary, *DNA vs. the Book of Mormon*, produced by Living Hope Ministries in Brigham City, Utah (*DNA* 2002). The ministry sits rather uneasily in a Mormon community and is keen to win converts at the margins. The documentary contrasts LDS beliefs with the findings of DNA research and contains interviews with Mormon and Gentile scientists. The Asian ancestry of Native Americans is stressed, but the data suggesting they arrived in the New World over 14,000 years ago is overlooked, perhaps in an attempt to make the video a little more palatable for those with fundamentalist beliefs.

LDS scholars with experience in DNA research have since joined the debate, although they have largely spoken only to Mormon audiences. In 2001, Scott Woodward, then Professor of Microbiology at BYU, addressed a Foundation for Apologetic Information and Research (FAIR) conference in Provo, Utah (Woodward 2001). In January 2003, Michael Whiting, Assistant Professor of Integrative Biology at BYU, gave a campus lecture entitled "Does DNA Evidence Refute the Authenticity of the Book of Mormon? Responding to the Critics" (Whiting 2003a). Well aware of the cool reception evolutionist concepts have usually received at BYU, Whiting could not help but note the peculiar irony that he, as a committed evolutionist, was coming to the defense of the Book of Mormon. A substantial amount of apologetic scholarship on the Lamanite-DNA question was published in late 2003 in the *Journal of Book of Mormon Studies* and *FARMS Review*. In a somewhat surprising development, links to the *Journal of Book of Mormon Studies* articles were provided on the official LDS website under the heading "Mistakes in the News":

> Recent attacks on the veracity of the Book of Mormon based on DNA evidence are ill considered. Nothing in the Book of Mormon precludes migration into the Americas by peoples of Asiatic origin. The scientific issues relating to DNA, however, are numerous and complex. Those interested in a more detailed analysis of those issues are referred to the resources below.
>
> The following are not official Church positions or statements. They are simply information resources from authors with expertise in this area that readers may find helpful. ... ("Mistakes" 2003)

Included on the LDS site was a link to an article written by Jeff Lindsay, a chemical engineer with no professional training in DNA research (Lindsay 2003).

APOLOGISTS RESPOND TO THE SCIENCE

In refuting claims that DNA findings challenge the historicity of the Book of Mormon, LDS scientists have advanced the idea that if one correctly understood the principles of population genetics, one would realize how improbable it would be to find evidence for an Israelite presence in the Americas (McClellan 2003, Whiting 2003b). Some LDS writers have tried to turn this to their advantage, claiming that the existence of people other than Book of Mormon colonists in pre-Columbian America is "supportive of careful readings of the Book of Mormon that have been available for many decades" (Peterson 2003a, Roper 2003a). The data, according to these writers, assist in overturning erroneous "long standing popular Mormon beliefs" (Roper 2003a) and eliminate the "parasitical and distorting" interpretations of church members who have been misguided (Peterson 2003a). LDS scholars have come to the conclusion that the Jaredite and Nephite peoples comprised a very small part of the vastly larger, pre-existing populations and that those who still believe the whole hemisphere was the stage for Book of Mormon events, or that substantial numbers of the literal descendants of Lehi live among contemporary natives of the Americas, are simply wrongheaded and have made "incorrect interpretation[s] of the text" (Peterson 2003a).

McClellan and Whiting, both assistant professors of integrative biology at BYU, have tried to nip the DNA controversy in the bud by retreating to a philosophical examination of the scientific method. According to McClellan, the Book of Mormon story line does not present a rejectable hypothesis (McClellan 2003). He repeats a mantra about what the limitations of science are—science proves "nothing"— while simultaneously acknowledging his belief that there was constant gene flow between Asia and the Americas. Whiting concedes that the different interpretations of the Book of Mormon, based either on the global or localized colonization schemes, imply different lineage histories and that the global hypothesis has been "falsified by current genetic evidence" (Whiting 2003b). D. Jeffrey Meldrum and Trent D. Stephens, two LDS biologists from Idaho State University, also confirm the unavoidable fact that most Native Americans are of Asian origin. Yet, the two hold out the possibility of the existence of a small subset of Middle Easterners living in the Americas which, they are confident, would be undetectable by science (Meldrum and Stephens 2003). While LDS scholars continue to split hairs, most admit, either overtly or subtly, that there is no "significant affirmative support" for the Book of Mormon (Peterson 2003b). They concede that Native Americans are principally of Asian ancestry.

Meldrum and Stephens gave the most objective summary of New World DNA research to date, informing LDS readers that 99.6 percent of Native Americans appear to have descended from Asian ancestors (Meldrum and Stephens 2003). Most who have spoken on the issue prefer to give a cursory description of the DNA research on Native Americans and then bury it within a more lengthy description of the problems and limitations of the methodology (Lindsay 2003, McClellan 2003, Roper 2003b, Woodward 2001, Whiting 2003a, 2003b). They like to argue that by using DNA technology, it would be virtually impossible to test the local colonization hypothesis, requiring that one discover remnants of a small Israelite family after its absorption into the larger indigenous population 2,600 years ago.

Suggesting that "we throw caution to the wind" and accept the local colonization hypothesis, Michael Whiting presented a list of problems—though not exhaustive—that he believed would frustrate any foolish DNA endeavor seeking to find a small colony of settlers in a limited area (Whiting 2003b). He underpinned his argument with a well-illustrated explanation of two closely related principles in population genetics—the "founder effect" and "genetic drift"—and then discussed how they would complicate the search for Lamanites. Briefly, the founder effect manifests itself in the restricted gene pool of a colonizing population because it is founded by such a limited number of individuals. Changes in the gene pool due to chance are referred to as genetic drift. These principles of population dynamics can complicate genetic studies, and Whiting illustrates the problems with the Lamanite genetic lineage. He presents three possible factors that could have produced an atypical, rather than representative, Middle Eastern genetic signal among the colonizers:

1. The original source population may not have had distinguishable genetic markers that would "unambiguously identify an individual as being from the Middle East."
2. The original colonizers may not have carried Middle Eastern genetic lineages even though they were from the Middle East.
3. The gene pool of the original colonizers may have changed over time by mutation or natural selection to the point that they no longer reflect the gene pool of the original source population.

All of these factors are theoretically possible, but it is not likely that they would completely frustrate the identification of Israelite DNA in the Americas. Middle Eastern populations are among the most heavily studied at the genetic level. These include populations that are geographically, culturally, and linguistically closely linked to Jewish populations. Mitochondrial and Y chromosome DNA studies have confirmed what has been known for some time, which is the close relationship between Jews and their neighbors and other Caucasian groups (see chapter 9). Given that the original Lehite/Mulekite founding parties originated from Jerusalem and that the Book of Mormon informs us that both groups were descended from Jewish ancestors, it would be hard to imagine that they were descendants of people from outside of the Middle East, let alone outside of Israel.

In his third point, Whiting implies that the genetic markers of the Amerisraelite founders could have become so altered that they would no longer be recognizable as Middle Eastern. David McClellan is similarly troubled and, after a lengthy and detailed introduction to molecular biology and population genetics, gives the problem a technical label: violations of the Hardy-Weinberg equilibrium (McClellan 2003). This theoretical principle is "blatantly" violated by the circumstances presented in the Book of Mormon, McClellan writes, because of (1) the migration of people in and out of their populations; (2) failure to maintain a constantly large population size; (3) non-random mating; and (4) natural selection. All of these would have affected the genetic structure of the Lehites and Mulekites.

However, if Book of Mormon people violate the Hardy-Weinberg equilibrium, they are not alone. All human populations violate the equilibrium in the same ways the hypothetical Lehites and Mulekites do, especially over a two-and-a-half-thousand-year period. This has not dissuaded scientists from using DNA methods to trace human ancestry over much greater time depths. The Hardy-Weinberg equilibrium is a theoretical model used to examine how populations change over time. If equilibrium is not maintained, it does not mean that genealogical ties become wholly unrecognizable. Mitochondrial and Y chromosome lineages are not likely to become altered to the point that they are no longer clear. In fact, in most cases, the mitochondrial DNA sequence remains exactly the same through numerous generations. The chance that a rare mutation will erase a single DNA spelling that defines a Middle Eastern lineage is approximately one in 16,500 since this is the number of DNA letters contained in mitochondrial DNA. Scientists typically use two or more DNA markers to characterize lineages, so the likelihood of a lineage being erased completely is extremely small. The Y chromosome lineages are even more robust. Just as branches in living trees converge at the trunk, the ancestors of Israelites living today will have lineages that meet at the Caucasian branch of the human family tree.

On the subject of genetic drift and founder effect, the apologists are quite correct. The gene pool arising from a small band of Israelite founders most likely would be different from its parental Middle Eastern population. We would expect some Middle Eastern DNA lineages to become more common and some to become less common purely as a result of chance. However, this would not significantly affect our ability to detect these lineages unless they became close to extinct. The apologists are also correct to assume that "swamping-out" of the colonizers by the introgression of genes from the native inhabitants would make it difficult to detect a Middle Eastern genetic signal (McClellan 2003, Whiting 2003a, Woodward 2001). But research on the Lemba of southern Africa suggests room for more optimism. In fact, John Sorenson previously used the Lemba to illustrate how, "when it is done right," DNA technology can link living people to ancient migrations of lost Israelites (Sorenson 2000b). The presence of Jewish Y chromosomes in the Lemba (see chapter 9) illustrates the persistence of the introduction of a small sample of Middle Eastern genes into a widely colonized foreign environment. Given that such a firm link could be found between Jews sharing a common ancestor 3,000 years ago (Thomas et al. 1998), it is odd that Sorenson and other LDS scientists would be so pessimistic about the chances of finding molecular signs of more recent Jewish ancestry in the Americas. The Book of Mormon informs us that the Lehite colonists "prospered" (2 Ne 5:11) and "multiplied exceedingly" (Jarom 1:8) soon after their arrival in their new homeland and that they became "exceedingly numerous" within a few hundred years (Omni 1:17).

Whiting suggests that another obstacle to detecting Lamanite lineages among Native American populations arises out of uncertainty about where the Lamanites might have been located or where their descendants might be (Whiting 2003b). He concedes, however, that church leaders have consistently associated Lamanites with the inhabitants of Central America. Since the chief geography apologist, Sorenson, and numerous others have identified Mesoamerica as the only possible candidate for the territory described in the Book of Mormon, it would be reasonable to examine the research that has been carried out among native tribes from this region.

In fact, the DNA lineages of Central America resemble those of other Native American tribes throughout the two continents. Over 99 percent of the lineages found among native groups from this region are clearly of Asian descent. Modern and ancient DNA samples tested from among the Maya generally fall into the major founding lineage classes (see appendix B). The Mayan Empire has been regarded by Mormons to be the closest to the people of the Book of Mormon because its people were literate and culturally sophisticated. However, leading New World anthropologists, including those specializing in the region, have found the Maya to be similarly related to Asians. Stephen L. Whittington, director of the Hudson Museum at the University of Maine and co-editor of *Bones of the Maya,* integrated dental, osteological, and molecular research on ancient Mayan skeletons to reconstruct Mayan culture and genetic relationships. Whittington was not aware of any scientists "in mainstream anthropology that are trying to prove a Hebrew origin of Native Americans" (*DNA* 2002). *Bones of the Maya* examines different time periods, social classes, and political systems from Mayan history and detailed research on skeletons buried 500 to 2,500 years ago at sites throughout Mesoamerica. According to Whittington, "Archaeologists and physical anthropologists have not found any evidence of Hebrew origins for the people of North, South and Central America" (*DNA* 2002).

The presence of the X lineage among Native Americans caught the apologist's interest. Initial studies failed to detect the X lineage in Asian populations, and there has been speculation about whether the X lineages now detected in Asia were directly ancestral to Native American lineages (Reidla et al. 2003) or were closer to the Middle East (Lindsay 2003, McClellan 2003). There are several problems associated with attempting to link the X lineage to Book of Mormon peoples. For instance, there is considerable genetic variation among New World X lineages (Brown et al. 1998), which would not be expected if they had arrived as recently as 2,600 years ago. The degree of variation implies that the X lineage was present in the first Americans over

14,000 years ago and accompanied the other four founding lineages across the ancient land bridge. The separation between New World and European X lineages is estimated to have taken place even farther back in time, about 22,000 years ago (Reidla et al. 2003). A further challenge is that the X lineage occurs in the highest frequencies in Alaska, Canada, and northeastern North America, a great distance from Mesoamerica, the widely accepted location for the Amerisraelite civilizations. The X lineage is generally absent in populations from Central America or Mesoamerica. There is also evidence that the X lineage is present in 4,000- to 7,000-year-old Native American remains, predating the Israelite groups mentioned in the Book of Mormon (Hauswirth et al. 1994, Ribeiro-dos-Santos et al. 1996).

By focusing on the detection of Middle Easterners in the New World, LDS writers have overlooked this key finding. Mitochondrial DNA research on over 7,000 individuals shows that about 99.6 percent of Native American lineages fall into the A-D and X lineage classes that originate from the migrations that occurred thousands of years before the Israelites came into existence. These lineages are not candidates for Israelite origin. Apologists are left with the remaining 0.4 percent of maternal lineages comprising European and African lineages. Since these are found at low frequency across North, Central, and South America and are much more common in tribes that interacted with early colonizers, it is reasonable to believe that these lineages are the result of admixture since Columbus. The pattern of appearance of these rare lineages is entirely consistent with an origin that is post-1492.

REVISIONIST BOOK OF MORMON SCHOLARSHIP

Currently, apologists defending the Book of Mormon assume a small Israelite group that encountered an overwhelmingly large existing population; they reject the previously defended global colonization hypothesis. If there had been a putative Israelite incursion into the New World, apologists have shrunk the potential impact of this event to the point that it is negligible. Given the current state of DNA affairs in the New World, Mormon scholars make the case that a thorough examination of the Book of Mormon shows that the impact of the Amerisraelites was so small that they barely mattered. An acceptance of the "local" hypothesis has necessitated the absorption of great numbers of Native Americans into the Israelite civilizations. As a consequence, an increasingly strenuous effort has been made to find Book of Mormon references to other people who have remained invisible to most Book of Mormon readers.

Sorenson and Roper have surveyed the major cultural, historical, and theological questions they think should be addressed before turning to DNA (Sorenson and Roper 2003), giving a description of various evidence they feel supports the arrival of transoceanic groups in the Americas prior to Columbus. They include among this evidence what they call "compelling" new research demonstrating Hebrew influence on the Uto-Aztecan language family in Central America and Utah. They reiterate Sorenson's earlier statements that the Nephites and Lamanites occupied all or part of Mesoamerica where the Book of Mormon history undoubtedly played out. Much of the paper is concerned with the significant "others" needed to be recruited into the Lehite civilizations to resolve population and geographical anomalies and to account for the Asian ancestry evident in DNA studies. Latter-day Saints "need to think of the Nephite record keepers as a minority—an elite minority at that—who, like most ruling minorities, tended to have their speech and customs eventually smothered by the speech and lifeways of the majority population" (Sorenson and Roper 2003).

We also now hear the argument that the term Lamanite soon lost its hereditary connotation and instead referred to a broad societal segment. It could be that the term referred to "all those" who were "led out of other countries by the hand of the Lord" (2 Ne 1:5; Sorenson and Roper 2003). Sorenson and Roper suggest that the "expression refers not only to the eventual Gentile (European) settlers of the 16th through 21st century but also to those ancient peoples whom the Lord brought as well." Apparently, Gentiles who inhabited the Americas before, during, and after the Book of Mormon period are potential Lamanites. Considering the dearth of genetic Lamanites, Meldrum and Stephens propose that the stain of having rejected God's covenants was passed from generation to generation via "memes," a term coined by Richard Dawkins in his book, *The Selfish Gene*, to denote a non-genetic unit of replication of human culture (Meldrum and Stephens 2003). Modern examples are melodies, icons, accents, and figures of speech that are inherited via mind-to-mind or face-to-face contact. Following this reasoning to a logical conclusion, it would not matter that a Native American had no genetic link to Israel; the fact that one of their ancestors came face to face with a genetic Lamanite would be sufficient for them to become Lamanites. Consequently, all Native Americans could be considered Lamanites, along with anyone else who has interacted with them.

Given such wide-ranging definitions, it would be helpful for LDS writers to clarify who would not fall under the Lamanite umbrella. According to Meldrum and Stephens, genetic traces of the House of Israel could be thought of as leaven in bread. Since too much leaven can be tasted in bread and decreases its quality, one should not expect to find genetic markers for the children of Lehi or even for the children of Abraham (Meldrum and Stephens 2003).

The expansion of the Lamanite family has facilitated the inclusion of non-Israelite "others" into the Lehite civilizations, which is important to the apologist's defense. However, most church members have not noticed references to other people in the Book of Mormon, either before the arrival of the Lehites or during the thousand-year account. Mormons have traditionally thought that any Asians who arrived in the New World did so after the Book of Mormon period drew to a close. Yet, Roper claims that the scriptural evidence for these others, even before the Lehites arrived, is "abundant," while evidence against the presence of others is "sparse and unimpressive" (Roper 2003a). There is much in the Book of Mormon that suggests otherwise.

For instance, the Nephites and Lamanites were to "be kept from all other nations, that they may possess this land unto themselves" (2 Ne 1:8-9). The prophecy of Lehi plainly spells out that other nations which would overrun the remnant Lamanite population, would only arrive after Columbus:

> But behold, when the time cometh that they shall dwindle in unbelief ... the judgments of him that is just shall rest upon them.
>
> Yea, he will bring other nations unto them, and he will give unto them power, and he will take away from them the lands of their possessions, and he will cause them to be scattered and smitten. (2 Ne 1:10-11)

Lehi's son Nephi saw visions of "many waters; and they divided the Gentiles from the seed of my brethren" (1 Ne 13:10). The Book of Mormon prophets specifically mention "other nations" in the latter days, while no explicit mention is made of any Gentiles living adjacent to the Nephites and Lamanites any time during their 1,000-year history. Roper dismisses references to an unpopulated New World, "kept from the knowledge of other nations," on the assumption that the Indian tribes "did not yet merit ... description [as] nations" (Roper 2003a).

The most compelling evidence that the author or people of the Book of Mormon did not envision other ethnic groups comes, not from what the book says, but from what was left unsaid. There are no explicit references to non-Israelites living with or adjacent to the Lehites or Jaredites, no "indigenous others" (Metcalfe 2004). In fact, over five hundred years after their arrival in the New World, the Lamanites "were a compound of Laman and Lemuel, and the sons of Ishmael, and all those who had dissented from the Nephites, who were Amalekites and Zoramites, and the descendants of the priest of Noah" (Alma 43:13). All of these groups traced their genealogies to Israel. One would have to assume that each of these groups, while silently assimilating with the surrounding Native American population, still retained the hereditary groupings conferred by the original founders.

When 8,000 Lamanites submit to baptism in about 30 BC, the Book of Mormon informs us, they do so because they have become convinced of the wickedness of the traditions of their fathers, meaning Laman and Lemuel. While a *Lamanite* may be someone other than a descendant of Laman, they are all described as being Israelites comprising specific designations as Lamanites, Lemuelites, and Ishmaelites (D&C 3:18). Nephite prophets repeatedly refer to the Lamanites as their "brethren" even after the climactic battle in AD 400 during which the Nephites are destroyed (Moroni 9:1). Nephite record keeper Moroni uses the term "my brethren" in reference to the army that had just exterminated the Nephites. It would be implausible to assume that any meaning other than genetic is implied.

Roper finds an abundance of lineage-related terms in the Book of Mormon and suspects this is why so many Mormons have assumed a genetic component to the Lehites and Mulekites (Roper 2003b). Familial terminology such as family, seed, children, descendants, fathers, sons, brethren, and so on usually suggest relatively clear genetic ties but are used in the Book of Mormon to indicate "ideological, social and political" links and groupings, according to Roper. The terms "descendant," "seed," and "children" are used in LDS theology to define a person's relationship with the theological House of Israel. People become the descendants, seed, or children of Abraham through adoption when they accept the gospel of Jesus Christ (Roper 2003b). This reasoning may be applied to the Nephites, where the text suggests that conversion implied an association with the Nephites, but adoption into the Lamanite nation was not a conversion experience. A plain reading of the Book of Mormon indicates that familial terms were meant to convey exactly what one would think they mean—a genetic link. Brent Metcalfe cites numerous post-Jaredite declarations of ancestry in the Book of Mormon (Metcalfe 2004) and notes that in every case, the speaker or writer goes on to trace his pedigree back to the founding Israelite immigrants. Metcalfe rejects the idea of a Nephite lineage history that was so ethnocentric that non-Israelites could remain invisible: "Such suggestions ... have no real explanatory power since *both* the Amerisraelites *and* the pre-Israelite Jaredites fail to mention indigenous 'others' and the Amerisraelite narrators exhibit no difficulty recognizing the Jaredites as non-Israelites who formerly inhabited the promised land" (Metcalfe 2004).

Acutely problematic arguments are required to explain away the numerous matter-of-fact statements by Joseph Smith and other church leaders who plainly taught that the Book of Mormon outlined the origins of the "Indians" in the western hemisphere. Statements by Joseph Smith that "the Indians were the literal descendants of Abraham" and that the gold plates gave "an account of the former inhabitants of this continent" are not as straightforward as they first appear, according to Roper. They do not claim to apply to *all* Native American history or ancestry (Roper 2003a).

Roper prizes apart a passage in Joseph Smith's letter to John Wentworth (see chapter 3) in which Smith makes this statement: "I was also informed concerning the aboriginal inhabitants of this country, and shown who they were, and from whence they came; a brief sketch of their origin, progress, civilization, laws, governments, of their righteousness and iniquity, and the blessings of God being finally withdrawn from them as a people was made known to me" (cf. Jessee 1984). In this quote, the words "brief sketch" are what caught the apologist's eye. Evidently, Smith was not privileged by God to receive a comprehensive knowledge of all American Indians but only regarding a portion of them (Roper 2003a). The apologist reminds us that these are not the actual words of Moroni but Joseph Smith's paraphrase of them. In the words of Roper, "There is no need to read into this statement any more than" Smith's opinion. Most church members would have trouble with this—stretching a prophet's words to defend a preconceived theory, but Roper has no problem submitting Smith to reinterpretation and investigation (Roper 2003b). Mormons tend to be hazy about what past leaders said exactly, preferring to focus on what the current, living prophets say. But the revisionist scholar sifts through statements from the present and past to find what suits his needs. As Metcalfe has written:

> [A]pologetic scholars ... have yet to explain cogently why all Book of Mormon characters—God included—seemingly know nothing about the hordes of indigenous peoples that the revisionist theories require; why Joseph Smith's revelation of the Book of Mormon is trustworthy enough to extract a detailed limited geography, yet his revelations about Amerindian identity and origins are flawed, if not erroneous; and why their word should count more than that of LDS prophets on the one hand, and that of secular scholars on the other. (Metcalfe 2004)

14

Moving the Spirit

People do not join the Church because of what they know. They join because of what they feel, what they see and want spiritually.

—Jeffrey R. Holland, 2001

Ten centuries ago a handful of Norse sailors slipped into Newfoundland, established small colonies, traded with local natives, then sailed back into the fog of history. In spite of the small scale of their settlements and the brevity of their stay, unequivocal evidence of their presence has been found, including metalwork, buildings, and Norse inscriptions. Just six centuries earlier, the Book of Mormon tells us, a climactic battle between fair-skinned Nephites and dark-skinned Lamanites ended a millennial dominion by a literate, Christian, Bronze Age civilization with a population numbering in the millions. Decades of serious and honest scholarship have failed to uncover credible evidence that these Book of Mormon civilizations ever existed. No Semitic languages, no Israelites speaking these languages, no wheeled chariots or horses to pull them, no swords or steel to make them. They remain a great civilization vanished without a trace, the people along with their genes.

Most Latter-day Saints have accepted the Book of Mormon based on what they *feel* about its message. Some now question the book because of what they *know* about its historical claims. Many are unsettled by the book's portrayal of a dark, corrupted race and the doctrine that America is God's promised land, issues that are uncomfortably reminiscent of the widespread prejudices of Joseph Smith's time. After decades of Mormon and non-Mormon academic research and LDS apologetics, the rank-and-file are beginning to find themselves faced, not only with the absence of reliable evidence for the presence of Israelites who sailed to America and established a great Christian civilization, but also the fact that these Israelites made no discernible contribution to the gene pool of native peoples, either on the continent or across the expanse of Polynesia. Many Latter-day Saints, discovering this for the first time, are disquieted by how far the Book of Mormon is from reality, as well as by how far the apologists have strayed from traditional Mormon beliefs.

The remaining option for many Mormons, whose integrity prevents them from ignoring the stark contradictions between science and Book of Mormon claims and who wish to maintain their fellowship in the church, is to keep their doubts to themselves. They are mindful of church president Gordon B. Hinckley's counsel to those who have problems with doctrine: "They can carry all the opinion they wish within their heads ... but if they begin to try to persuade others, then they may be called in to a disciplinary council" (Hinckley 1998). Consequently, there are many bright and faithful Mormons who have chosen to keep their intellectual struggles between their own two ears. Increasingly, these Saints occupy the ranks of the less active or the back rows of LDS congregations. Many in Utah choose this option because to do otherwise risks alienation from community, work environment, and families. In communities where workmates, closest friends, and most relatives are multi-generational Latter-day Saints, Mormonism is a part of one's cultural heritage and identity.

A growing number of Mormons find comfort in the church's current outreach to people of other faiths and attempt to establish a position of respect and acceptance within the larger society. This includes, in part, a willingness to emphasize similarities and to de-emphasize some of the differences, among which the claim to have a real history of ancient America may one day be counted (Metcalfe 1993). Some scholars within the church have decided that the once all-consuming battle over the Book of Mormon's historicity was a waste of energy and that members should have devoted their attention to mining the book's spiritual gems (Thomas 1999). The Community of Christ, formerly the Reorganized Church of Jesus Christ of Latter Day Saints, tolerates a range of opinions concerning the Book of Mormon. This Missouri-based church is the only sizeable group thus far to have broken away from the Utah church. Its worldwide membership is about 250,000. The Community of Christ discarded the problematic Book of Abraham when the papyri Joseph Smith used to "translate" the record were discovered and studied. Some of the Missouri church's senior leadership consider the Book of Mormon to be inspired historical fiction. For leaders of the Utah church, this is still out of the question. The Brethren, and most Mormons, believe that the historical authenticity of the Book of Mormon is what shores up Joseph Smith's prophetic calling and the divine authenticity of the Utah church. The Brethren undoubtedly also link the flagging numerical performance of the Community of Christ with this and other dramatic doctrinal about-faces such as the decision to ordain women to the priesthood.

The Utah leadership remains unwilling or incapable of respectfully approaching those who honestly struggle with the disparity between the Book of Mormon's claims and the documented history of Native Americans and Polynesians. With its own schools, universities, public relations, and an annual income of about six billion U.S. dollars per year (Ostling and Ostling 1999), the LDS church has the resources necessary to defend its stand. With employment in LDS educational institutions and the Church Educational System conditioned on unquestioning obedience to leaders, the church receives a continuous influx of new apologetic muscle, which is required to shore up the flagging theology. But just how far can that muscle be stretched in order to save the Book of Mormon?

COURTING THE ENEMY

Most LDS apologists now accept that Native Americans are principally descended from Siberian ancestors who migrated across the Bering Strait thousands of years before Lehi arrived and that the descendants of Lehi made up an infinitesimally smaller proportion of the New World populations. However, this change in perspective has not been granted the church's blessing in any official way. The general membership would not believe that Lehi's descendants could have made such a minimal impact in the Americas. In fact, millions of Mormons consider Lehi to be the father of the New World and believe that he stands at the head of their own family pedigrees. Despite decades of work by apologists, their work has yet to be discussed openly in the various public forums the church sponsors. In attempting to serve two masters, the apologists may end up pleasing none. Ordinary church members will resist the idea that Nephite and Lamanite civilizations were powered by displaced Siberians who fought their wars and erected their temples for them when they were not granted the dignity of a single mention in the record. The Brethren will find such stark rationalizations equally unpalatable, particularly if broadcast to wide audiences because the ideas contrast so profoundly with the deeply held beliefs of most Mormons.

There is the further problem for apologists that in trying to rescue the Book of Mormon from science, they have had to reject the clear pronouncements of every church president from Joseph Smith to the present. While apologists have long accepted the fact that other groups outside of the Book of Mormon record made their way to the New World, few apologists would have predicted that the Lamanite influence would be virtually undetectable. The accumulating DNA data have provided the first quantitative measure of an Israelite presence in the New World gene pool, and it is slim to none. The apologists are unable to find an Israelite genetic signature in the Pacific, the Americas, or in the more limited territories of Central America and Mesoamerica outside of what can be explained by recent migration. So they have chosen instead to reinterpret the meaning of statements by modern prophets.

Acceptance of the molecular evidence creates further problems for Latter-day Saints. Anyone reading the responses coming from LDS biologists will discover that they have not quibbled with the evidence for the colonization of the Americas over 13,000 years ago, for the occupation of Asia and Australia roughly 60,000 years ago, and for the emergence of humans in Africa over 100,000 years ago. Church members who were initially only curious about the Israelite DNA issue are confronted by challenges to other closely held beliefs such as the placement of Adam and Eve on the earth and a post-Flood colonization, events that most Mormons believe occurred within the last six thousand years. LDS doctrine clearly states that Adam and Eve lived in the vicinity of Independence, Missouri, despite abundant evidence that all of the earliest members of the human family dwelt in sub-Saharan Africa. LDS apologists need to explain how Australian Aboriginals and Native Americans and many other native groups have lived continuously on separate hemispheres for tens of thousands of years unperturbed by a global deluge.

REINTERPRETING SCRIPTURE

The research at BYU by defenders of the Book of Mormon has produced a sweeping reinterpretation of LDS doctrine. BYU professors have felt compelled to shrink the scale of the assumed Israelite incursion into the Americas and to magnify the term Lamanite to embrace numerous Native Americans who inhabited the continent for over 10,000 years, contrary to a plain reading of the Book of Mormon.

In the past, the Brethren, along with the rank and file, have not taken well to reinterpretations of the Book of Mormon, especially when the revisionists have impinged on the scale and location of the Lehite presence. Already in 1938 apostle and future church president Joseph Fielding Smith spoke out against advocates of a limited geography. Scholars who have argued for a Mesoamerican Cumorah have been greeted with mixed suspicion and hostility (Reynolds 1999). In 1985 the *Ensign* published a condensation of John Sorenson's trend-setting work, *An Ancient American Setting for the Book of Mormon*. However, Sorenson's work found acceptance only among a small circle of LDS intellectuals. The research at BYU remains virtually unknown by most Mormons, who prefer to see the Lehite setting as a literal reality and global in scale. They strongly resist factual arguments to the contrary.

The reason for this is that Latter-day Saints have been taught such concepts from childhood—anecdotal stories and popular mythology about the Indians and Polynesians. A favorite children's hymn, sung to the beat of a stereotypical Indian tom-tom dance, is "Book of Mormon Stories that My Teacher Tells to Me." Generations of youth have been assured that scientific institutions like the Smithsonian have used the Book of Mormon to guide their research. Many thousands of church members, young and old, have made a pilgrimage or two to the Hill Cumorah Pageant in New York to see a clichéd Book of Mormon extravaganza brimming with white skinned, sword-wielding Nephites and half naked, dark-skinned, stereotypical Lamanite "Indians." It would be extremely painful for people to escape the heavy historical baggage accompanying this folklore.

Attempts to redefine the term Lamanite so that it includes superior numbers of non-Israelite Native Americans make for stimulating apologetics, but it will remain extra-canonical because the Book of Mormon is silent on the matter. The book's narrative is uncomplicated on that score. It clearly depicts the settlement of an area of the world that was previously unpopulated. It portrays Lehi and his posterity, isolated from all other nations on earth, prospering to the extent that they built a great civilization. The descendants of Laman remained alive to see the arrival of the Gentiles whose coming was predicted and connected to the recovery of the gold plates.

Rather than side with the revisionists, the General Authorities continue to tell members in selected areas of the world that they are the offspring of Father Lehi, a doctrine Joseph Smith and all subsequent prophets have endorsed. As recently as 2000, the First Presidency reminded Saints in Central and South America that they are the "sons and daughters of Father Lehi." Further apostolic authority was added when these statements were incorporated into temple dedicatory prayers. Hundreds of thousands of people have heard this doctrine preached from the mouths of prophets, and many people consider themselves to be card-carrying Lamanites, having received a patriarchal blessing informing them of their Israelite ancestry via Lehi. On the basis of this belief, the church invests a disproportionate amount of funding for schools in Central and South America and Polynesia compared to what it spends in Africa or Asia, for instance. The Mormon church continues to impose a deeply flawed colonial mythology upon native peoples across the Americas and the Pacific, with apparent disregard for these peoples' own local mythologies, cultural history, and distinct genealogies.

FEELINGS OVER FACTS

It is likely that the church will continue to carry on a rearguard defense of the Book of Mormon, quietly promoting apologetic scholarship but not officially endorsing it. Officially the church claims to have no position regarding Book of Mormon geography, yet it provides funds for researchers whose focus is almost exclusively on Mesoamerica. Publicly, the church urges members to steer clear of any attempt to link the Book of Mormon with current geographical locations. Meanwhile, apologists subject the Book of Mormon to radical reinterpretation to accommodate their findings. Their apostolic patrons know to distance themselves from these interpretations, at least publicly, so the church has the option, after seeing how members might react to the gradual accumulation of information and apologetic responses, to either endorse or denounce them as the timing might dictate.

It seems that among the obstacles facing the church, the real stumbling block is not the failure to find evidence for horses, metallurgy, or the wheel in the New World, or the fact that there is no evidence for a Hebrew influence in Mesoamerica, or the preponderance of Asian DNA among living Native Americans and Polynesians. The real challenge comes from a failure to openly confront the evidence and state what it means for the church, as well as a failure to accommodate the apologists, who themselves feel hemmed in by the church's insistence that members believe tenets that are clearly untrue. From the perspective of some members, the apologists have been lured into the enemy camp of mainstream science. The theories of the apologists concerning a miniscule Lehite colony that existed in some unknown corner and had no lasting impact on the Americas are equally unsatisfying to mainstream scientists. Orthodox Mormons cannot conceive of such a reinterpretation of the Book of Mormon, and therefore the current prophets are reluctant to publicly address the problems. This leaves the apologists cut off from the broader church community.

The Brethren no doubt recognize that to change the way Mormons think about the Book of Mormon would bring disruption and turmoil and risk undermining the foundation on which many people have based their religious convictions. The faith of most Latter-day Saints is anchored by an emotional, feeling-based "knowledge" that the Book of Mormon is a true history of the Americas—a knowledge that is unencumbered by the facts. These religious feelings are believed to be the means by which the Holy Spirit can reveal all truth to all mankind on earth. Millions of members feel a familial bond with Father Lehi, an emotion that frequently plays a central role in people's conversion to the church. The General Authorities are aware of just how deep-seated and crucial these feelings are in the processes of conversion and retention. To date, no workable middle ground has been discovered—no means to reinterpret the Book of Mormon in a way that would detach it from the real histories of Native Americans and Polynesians without doing damage to everything else the church professes on spiritual and moral issues. The church's dim view of scientists and intellectuals extends to theologians and philosophers, who may be the best suited to render assistance at such a time. Failing that, the conflict promises to continue into the very distant future without hope of resolution.

Appendixes

A

Testing for Mitochondrial DNA Lineages

Two methods of DNA analysis are commonly used to determine the mitochondrial DNA lineage of an individual. These are (1) restriction site analysis and (2) DNA sequencing of a short region of the mitochondrial DNA known as the control region or D-loop. Restriction analysis is performed by digesting the mitochondrial DNA with restriction enzymes that recognize specific sequences in the DNA. The loss or gain of the ability of a particular restriction enzyme to cut DNA at a site indicates a specific change in the DNA spelling. DNA sequencing, on the other hand, delivers the order of DNA letters in a particular piece of DNA, clearly revealing the alterations in DNA spelling.

The founding lineages, or lineages present in the first Americans, have specific DNA spellings that can be detected by either method. These changes are usually defined with reference to a mitochondrial sequence known as the Cambridge Reference Sequence, one of a number of lineages that belong to the H lineage family, the most common lineage group found among Caucasians.

1. Restriction Site Analysis

The following table lists DNA restriction sites that define the five major founding lineages in American Indians. The numbers refer to DNA base positions in the 16,569 bases of the mitochondrial genome

(Anderson et al. 1981). The plus and minus signs indicate the gain or loss of a restriction site.

Lineage	*Hae*III (bp 663)	9bp deletion (bp 8271-81)	*Alu*I (bp 13262)	*Alu*I (bp 5176)	*Dde*I (bp 1715)
Cambridge (H)	–	–	+	+	+
A	+	–	+	+	+
B	–	+	+	+	+
C	–	–	–	+	+
D	–	–	+	–	+
X	–	–	+	+	–

Source: Schurr et al., *American Journal of Human Genetics* 46 (1990): 613-23; Kaestle and Smith, *American Journal of Physical Anthropology* 115 (2001): 1-12.

2. Control Region Sequencing

Changes in Native American mitochondrial DNA control region sequences in relation to the Cambridge Reference Sequence. Dots indicate bases identical to the Cambridge sequence.

Lineage	Positions in the Cambridge DNA sequence (16, xxx)										
Cambridge	111	189	217	223	278	290	298	319	325	327	362
	(C)	(T)	(T)	(C)	(C)	(C)	(T)	(G)	(T)	(C)	(T)
A	T	•	•	T	•	T	•	A	•	•	C
B	•	C	C	•	•	•	•	•	•	•	•
C	•	•	•	T	•	•	C	•	C	T	•
D	•	•	•	T	•	•	•	•	C	•	C
X	•	C	•	T	T	•	•	•	•	•	•

Source: Schurr et al., *American Journal of Human Genetics* 46 (1990): 613-23; Brown et al., *American Journal of Human Genetics* 63 (1998): 1852-61.

B

Maternal DNA Lineages in the New World

1. New World Summary

	Maternal Lineage (number of individuals)						
Population	A	B	C	D	X†	Other	Total
Alaska	288	4	13	379	0	24	708
Greenland	82	0	0	0	0	0	82
Canada	443	42	82	29	55	3	654
United States*	554	633	379	185	61	16	1,828
Central America**	291	117	77	22	0	4	511
South America	676	1,175	914	683	0	86	3,534
Total	2,334	1,971	1,465	1,298	116	133	7,317
Percentage	32%	27%	20%	18%	2%	2%	100%

Source: See source references for tables 3-6. †X lineage numbers are estimates based on Table 2 in this appendix. *This category refers to the continental United States. **Central America includes the Caribbean.

2. Lineage X and other New World Maternal Lineages

Lineage Category	Number	Percentage
U.S. & Canadian X (estimate)[a]	116	1.6
Ecuador-Cayapa[c]	26	0.4
European and African (estimate)[d]	32	0.4
Alaskan (not fully characterized)[b]	24	0.3
South America (not fully characterized)[e]	51	0.7
Total	**249**	**3.4**

[a] See Smith et al., *American Journal of Physical Anthropology* 110 (1999): 271-84.

[b] Maternal lineages of Alaskan populations are currently not fully characterized. Some may be X lineages and some may be additional Asian lineages.

[c] The Cayapa lineage—found exclusively among the Cayapa Indians of Ecuador—lacks mutations characteristic of the A, B, C, D, or X lineages. However, it contains other mutations that indicate it originated in Asia (see Rickards et al., *American Journal of Human Genetics* 65 [1999]: 519-30).

[d] See Smith et al., *American Journal of Physical Anthropology* 110 (1999): 271-84.

[e] These South American lineages are insufficiently characterized to assign them to either of the five founding lineage groups.

3. Alaska, Greenland, and Canada

	Maternal Lineage					
Population	A	B	C	D	X+Other	Sources[††]
Alaska						
Old Harbor Eskimo* [115]	71	4	0	40	0	5
Ouzinkie Eskimo* [41]	30	0	2	6	3	5
Gambell Eskimo* [50]	29	0	7	13	1	5
Yupik Eskimo* [165]	43	0	3	106	13	13
Savoonga Eskimo* [49]	46	0	0	1	2	5
St. Paul Aleut* [72]	18	0	1	48	5	5

Alaska, Greenland, and Canada (continued)

Population	A	B	C	D	X+Other	Sources[††]
Aleutian Islands* [179]	51	0	0	128	0	60
Commander Islands Aleut* [37]	0	0	0	37	0	36
Greenland						
Greenland Eskimo* [82]	82	0	0	0	0	37
Canada						
Dogrib† [154]	140	0	3	0	11	5
Inuit* [30]	29	0	0	1	0	10
Haida† [63]	59	0	3	1	0	3, 6
Bella Coola [57]	40	4	5	7	1	3, 6
Nuu-Chah-Nulth [77]	34	3	14	17	9	3, 7, 30
Chippewa [15]	4	2	5	0	4	3
Mohawk [128]	71	21	29	2	5	13
Ojibwa [43]	22	3	7	0	11	3
Nth Ontario Ojibwa [26]	17	1	2	0	6	40
Turtle Mountain Chippewa [28]	16	5	5	0	2	41
Manitoulin Island Ojibwa [33]	11	3	9	1	9	40
Total [1,444]	813	46	95	408	82	

*Eskaleut linguistic group. †NaDene linguistic group. All other populations listed belong to the Amerind linguistic group (see Greenberg et al., *Current Anthropology* 27 [1986]: 477-97).[††] The source references for tables 3-6 are listed at the end of appendix B. The number in brackets after each population name is the number of individuals tested.

4. Continental United States

	Maternal Lineage					
Population	A	B	C	D	X+Other	Sources
Oneota# [108]	34	13	46	9	6	9
Adena# [16]	2	7	1	5	1	55
Illinois Hopewell# [47]	16	6	16	8	1	55
Ohio Hopewell# [34]	14	3	10	7	0	59

Continental United States (continued)

Population	A	B	C	D	X+Other	Sources
Fort Ancient# [18]	5	2	9	2	0	55
Michigan Woodland# [16]	6	0	6	2	2	57
Oklahoma Muskoke# [71]	26	11	7	27	0	5
Navajo† [204]	98	87	13	2	4	3, 10, 42
Apache† [209]	143	40	20	2	4	3, 10, 42
Wisconsin Chippewa [62]	17	3	22	2	18	41
Oklahoma Cherokee [19]	4	4	10	1	0	10, 41
Stillwell Cherokee [37]	4	17	16	0	0	10, 41
Sioux [45]	25	9	8	2	1	10, 41, 43
Chippewa/Kickapoo [62]	30	7	12	0	13	10
Cheyenne/Arapahoe [35]	12	4	12	5	2	10, 41
Pawnee [5]	2	3	0	0	0	10, 41
Yakima [42]	4	26	3	6	3	21
Micmac/Narragansett [7]	2	0	1	0	4	10
Cherokee [16]	0	5	5	0	6	10
Siouan [34]	18	6	5	2	3	10
California Penutian [17]	2	7	1	7	0	10
Chickasaw [8]	1	6	1	0	0	10, 50
Choctaw [27]	20	5	1	1	0	10, 50
Creek [39]	14	6	8	11	0	10, 50
Seminole [40]	25	10	3	2	0	50, 51
Zuni [26]	4	20	2	0	0	10, 53
Washo [28]	0	15	10	3	0	10
Akimel O'odham [43]	2	23	17	0	1	53
Tohono O'odham [37]	0	21	14	2	0	53
Yuman [72]	2	45	25	0	0	10, 53, 54
Kiliwa [7]	0	7	0	0	0	10, 53, 54
Seri [8]	0	1	7	0	0	53
Quechan/Cocopa [23]	0	15	7	0	1	10
Havasupai/Hualapai/Yavapai/Mojave [18]	2	9	7	0	0	10

Continental United States (continued)

Population	A	B	C	D	X+Other	Sources
Kiliwa/Paipai [11]	0	10	1	0	0	11
Kumiai [16]	0	10	6	0	0	10
Cochimi [13]	1	6	6	0	0	10
Salinan/Chumash [11]	5	2	1	3	0	10
Jemez [36]	0	31	1	1	3	10
Paiute/Shoshoni [94]	0	40	9	45	0	45, 53
Pima [30]	2	15	13	0	0	3, 11
California Uto-Aztecan [14]	0	4	6	4	0	10
Fremont, Great Salt Lake# [30]	0	22	4	2	2	25
Anasazi# [54]	9	36	7	0	2	44, 52
Pyramid Lake# [18]	2	6	0	10	0	45
Stillwater Marsh# [21]	1	8	0	12	0	45
Total [1,828]	554	633	379	185	77	

DNA tests conducted on ancient remains.

† NaDene linguistic group.

5. Central America / Caribbean

	Maternal Lineage					
Population	A	B	C	D	X+Other	Sources
Nahua/Cora [32]	17	11	2	0	2	10
Alta Mixtec [15]	11	2	2	0	0	8
Baja Mixtec [14]	13	1	0	0	0	8
Mixe [16]	10	5	1	0	0	8
Zapotec [15]	5	5	5	0	0	8
Maya (Mexico) [27]	14	6	4	2	1	3, 11
Maya# [25]	21	1	2	0	1	46
Maya (Copan)# [8]	0	0	8	0	0	28
Bribri/Cabecar [24]	13	11	0	0	0	3
Boruca [14]	3	10	0	1	0	3

Central America / Caribbean (continued)

Population	A	B	C	D	X+Other	Sources
Guatuso [20]	17	3	0	0	0	12
Teribe [20]	16	4	0	0	0	12
Guyami [16]	11	5	0	0	0	3
Huetar [26]	18	1	0	7	0	14
Ngobe [46]	31	15	0	0	0	15
Kuna [79]	61	18	0	0	0	3, 16
Embera [44]	23	10	11	0	0	20
Wounan [31]	6	9	15	1	0	20
Tainos (Caribbean)# [24]	0	0	18	6	0	1
Ciboneys (Cuba)# [15]	1	0	9	5	0	49
Total [511]	291	117	77	22	4	

DNA tests were conducted on ancient remains.

6. South America

	Maternal Lineage					
Population	A	B	C	D	X+Other	Sources
Columbian, > 26 tribes [701]	223	212	199	46	21	17, 56
Columbian mummies# [6]	3	1	2	0	0	22
Embera [22]	16	5	0	0	1	35
Kogui [50]	29	0	21	0	0	48
Arsario [50]	32	0	18	0	0	48
Ijka [40]	37	1	2	0	0	48
Ingano [27]	4	12	10	0	1	35
Wayuu [92]	26	26	37	0	3	35, 48
Zenu [37]	7	15	11	2	2	35
Bolivian [233]	51	121	48	7	6	2
Cayapa [120]	35	48	11	0	26	39
Piaroa [10]	5	0	1	4	0	3
Makiritari [10]	2	0	7	1	0	3

South America (continued)

Population	A	B	C	D	X+Other	Sources
Brazilian Amazon, 8 tribes [139]	41	39	38	19	2	23
Brazilian Amazon# [18]	5	1	4	1	7	24
Arara [9]	1	2	6	0	0	26
Wayampi [21]	17	4	0	0	0	26
Kayapo [13]	6	7	0	0	0	26
Xavante [25]	4	21	0	0	0	4
Zoro [30]	6	2	4	18	0	4
Gaviao [27]	4	4	0	19	0	4
Yanomamo [262]	0	96	124	42	0	3, 18, 58
Macushi [10]	1	2	3	4	0	3
Marubo [10]	1	0	6	3	0	3
Ticuna [82]	12	8	30	32	0	5, 11, 35
Wapishana [12]	0	3	1	8	0	3
Mataco [129]	11	70	11	35	2	3, 32, 33
Toba [56]	11	23	3	17	2	32
Pilaga [41]	2	15	11	12	1	32
Chorote [34]	5	15	8	6	0	32, 33, 38
Kraho [14]	4	8	2	0	0	3
Quechua [19]	5	7	1	6	0	5
Aymara [172]	11	116	21	24	0	5
Atacemeno [63]	9	45	6	3	0	5, 31
Argentinian Mapuche [97]	9	34	20	28	6	19, 31
Arequipa [22]	2	15	3	2	0	47
Tayacaja [61]	13	20	8	18	2	47
San Martin [22]	2	12	1	6	1	47
Punenos [65]	8	42	11	4	0	29
Chilean Mapuche [156]	2	18	67	69	0	17, 38
Peheunche [100]	2	9	37	52	0	5
Huilliche [118]	5	34	22	57	0	5, 31
Aonikenk# [19]	0	0	6	13	0	27, 61
Yamana# [17]	0	0	15	2	0	27, 61

South America (continued)

Population	A	B	C	D	X+Other	Sources
Kaweskar# [30]	0	0	7	23	0	27, 61
Selknam# [16]	0	0	9	6	1	27, 61
Pehuenche (Trapa-Trapa) [105]	3	11	43	48	0	38
Yaghan [21]	0	0	10	11	0	34
Whichi [72]	4	45	2	19	2	34
Tehuelche [29]	0	6	7	16	0	38
Total [3,534]	676	1,175	914	683	86	

DNA tests were conducted on ancient remains.

SOURCES

1. Lalueza-Fox et al., *Annals of Human Genetics* 65 (2001): 137-51.
2. Bert et al., *Human Biology* 73 (2001): 1-16.
3. Torroni et al., *American Journal of Human Genetics* 53 (1993): 563-90.
4. Ward et al., *American Journal of Human Biology* 8 (1996): 317-23.
5. Merriwether et al., *American Journal of Physical Anthropology* 98 (1995): 411-30.
6. Ward et al., *Proceedings of the National Academy of Science* 90 (1993): 10663-67.
7. Ward et al., *Proceedings of the National Academy of Science* 88 (1991): 8720-24.
8. Torroni et al., *American Journal of Human Genetics* 54 (1994): 303-18.
9. Stone & Stoneking, *Philosophical Transactions of the Royal Society* 354 (1999): 153-59.
10. Lorenz & Smith, *American Journal of Physical Anthropology* 101 (1996): 307-23.
11. Schurr et al., *American Journal of Human Genetics* 46 (1990): 613-23.
12. Torroni et al., *Proceedings of the National Academy of Science* 91 (1994): 1158-62.
13. Merriwether & Ferrell, *Molecular Phylogenetic and Evolution* 5 (1996): 241-46.
14. Santos et al., *Human Biology* 66 (1994): 963-77.
15. Kolman et al., *Genetics* 140 (1995): 275-83.
16. Batista et al., *Human Molecular Genetics* 4 (1995): 921-29.
17. Horai et al., *Molecular Biology and Evolution* 10 (1993): 23-47.
18. Easton et al., *American Journal of Human Genetics* 59 (1996): 213-25.
19. Ginther et al., in *DNA Fingerprinting: State of the Science,* eds. Pena et al. (Boston: Birkhauser, 1993), 211-19.

20. Kolman & Bermingham, *Genetics* 147 (1997): 1289-1302.
21. Shields et al., *American Journal of Human Genetics* 53 (1993): 549-62.
22. Monsalve et al., *Annals of Human Genetics* 60 (1996): 293-303.
23. Santos et al., *Annals of Human Genetics* 60 (1996): 305-19.
24. Ribeiro dos Santos et al., *American Journal of Physical Anthropology* 101 (1996): 29-37.
25. Parr et al., *American Journal of Physical Anthropology* 99 (1996): 507-18.
26. Bortolini et al., *Annals of Human Genetics* 62 (1998): 133-45.
27. Lalueza et al., *Human Molecular Genetics* 6 (1997): 41-46.
28. Merriwether et al., *Experientia* 50 (1994): 592-601.
29. Dipierri et al., *Human Biology* 70 (1998): 1001-10.
30. Forster et al., *American Journal of Human Genetics* 59 (1996): 935-45.
31. Bailliet et al., *American Journal of Human Genetics* 55 (1994): 27-33.
32. Demarchi et al., *American Journal of Physical Anthropology* 115 (2001): 199-203.
33. Bianchi et al., *Brazilian Journal of Genetics* 18 (1995): 661-68.
34. Bravi et al. (1995), cited in Moraga et al., *American Journal of Physical Anthropology* 113 (2000): 19-29.
35. Mesa et al., *American Journal of Human Genetics* 67 (2000): 1277-86.
36. Derbeneva et al., *American Journal of Human Genetics* 71 (2002): 415-21.
37. Saillard et al., *American Journal of Human Genetics* 67 (2000): 718-26.
38. Moraga et al., *American Journal of Physical Anthropology* 113 (2000): 19-29.
39. Rickards et al., *American Journal of Human Genetics* 65 (1999): 519-30.
40. Scozzari et al., *American Journal of Human Genetics* 60 (1997): 241-44.
41. Malhi et al., *Human Biology* 73 (2001): 17-55.
42. Budowle et al., *International Journal of Legal Medicine* 116 (2002): 212-15.
43. Smith et al., *American Journal of Physical Anthropology* 110 (1999): 271-84.
44. Carlyle et al., *American Journal of Physical Anthropology* 113 (2000): 85-101.
45. Kaestle and Smith, *American Journal of Physical Anthropology* 115 (2001): 1-12.
46. Gonzalez-Oliver et al., *American Journal of Physical Anthropology* 11 (2001): 230-35.
47. Fuselli et al., *Molecular Biology and Evolution* 20 (2003): 1682-91.
48. Briceno et al. (2003), visual display at the International Genetics Congress, Melbourne.
49. Lalueza-Fox et al., *American Journal of Physical Anthropology* 121 (2003): 97-108.
50. Bolnick & Smith, *American Journal of Physical Anthropology* 122 (2003): 336-54.
51. Huoponen et al., *European Journal of Human Genetics* 5 (1997): 25-34
52. Carlyle & O'Rourke (2001), visual display at the Seventieth Meeting of the American Association of Physical Anthropologists.
53. Malhi et al., *American Journal of Physical Anthropology* 120 (2003): 108-24.
54. Smith et al., *American Journal of Physical Anthropology* 111 (2000): 557-72.
55. Bolnick (2003), visual display at the Seventy-second Meeting of the American Association of Physical Anthropologists.

56. Keyeux et al., *Human Biology* 74 (2002): 211-33.
57. Grennan & Merriwether (2003), visual display at the Seventy-second Meeting of the American Association of Physical Anthropologists.
58. Williams et al., *American Journal of Physical Anthropology* 117 (2002): 246-59.
59. Mills, Ph.D. diss., Ohio State University, 2003.
60. Rubicz et al., *Human Biology* 75 (2003): 809-35.
61. Garcia-Bour et al., *American Journal of Physical Anthropology* 123 (2004): 361-70.

C

Statement from the Smithsonian Institution

Department of Anthropology, Smithsonian Institution, 1996

1. The Smithsonian Institution has never used the Book of Mormon in any way as a scientific guide. The Smithsonian archaeologists see no direct connection between archaeology of the New World and the subject matter of the book.

2. The physical type of American Indian is basically Mongoloid, being most closely related to that of the peoples of eastern, central and northeastern Asia. Archaeological evidence indicates that the ancestors of the present Indians came into the New World—probably over a land bridge known to have existed in the Bering Strait region during the last Ice Age—in a continuing series of small migrations beginning from about 25,000 to 30,000 years ago.

3. Present evidence indicates that the first people to reach this continent from the East were the Norsemen who briefly visited the northeastern part of North America around A.D. 1000 and then settled in Greenland. There is nothing to show that they reached Mexico or Central America.

4. One of the main lines of evidence supporting the scientific finding that contacts with Old World civilizations, if indeed they occurred at all, were of very little significance for the development of American Indian civilizations is the fact that none of the principal Old World domesticated food plants or animals (except the dog) occurred in the New World in

pre-Columbian times. American Indians had no wheat, barley, oats, millet, rice, cattle, pigs, chickens, horses, donkeys, [or] camels before 1492 (camels and horses were in the Americas, along with the bison, mammoth, and mastodon, but all these animals became extinct around 10,000 B.C. at the time the early big game hunters spread across the Americas).

5. Iron, steel, glass, and silk were not used in the New World before 1492 (except for occasional use of unsmelted meteoric iron). Native copper was worked in various locations in pre-Columbian times, but true metallurgy was limited to southern Mexico and the Andean region, where its occurrence in late prehistoric times involved gold, silver, copper, and their alloys, but not iron.

6. There is a possibility that the spread of cultural traits across the Pacific to Mesoamerica and the northwestern coast of South America began several hundred years before the Christian era. However, any such interhemispheric contacts appear to have been the results of accidental voyages originating in eastern and southern Asia. It is by no means certain that even such contacts occurred with the ancient Egyptians, Hebrews, or other peoples of Western Asia and the Near East.

7. No reputable Egyptologist or other specialist on Old World archaeology, and no expert on New World prehistory, has discovered or confirmed any relationship between archaeological remains in Mexico and archaeological remains in Egypt.

8. Reports of findings of ancient Egyptian, Hebrew, and other Old World writings in the New World in pre-Columbian contexts have frequently appeared in newspapers, magazines and sensational books. None of these claims has stood up to examination by reputable scholars. No inscriptions using Old World forms of writing have been shown to have occurred in any part of the Americas before 1492 except for a few Norse rune stones which have been found in Greenland.

9. There are copies of the Book of Mormon in the library of the National Museum of Natural History, Smithsonian Institution.

D

Websites on Mormon Topics

Denominational

http://www.lds.org

The official website of the Church of Jesus Christ of Latter-day Saints. It contains information for members and nonmembers including a description of the basic beliefs and recent news of the church. Other resources include media information and a searchable online gospel library containing the four LDS scriptures and church magazines.

http://www.cofchrist.org

Official website of the Community of Christ, formerly known as the Reorganized Church of Jesus Christ of Latter Day Saints, the name having been changed on 6 April 2001.

Educational

http://www.byu.edu

Official website of Brigham Young University, the largest private university in the United States, founded in 1875. Its current enrollment is approximately 33,000 students across a broad range of disciplines.

http://www.byui.edu

Official website of the Idaho campus of Brigham Young University, formerly known as Ricks College. Located in Rexburg, Idaho, the campus has about 10,000 students.

http://www.byuh.edu

Official website of the Hawaii campus of Brigham Young University. Established in 1955, the campus has about 2,200 students studying liberal arts and professional programs in business and education.

LDS Genealogy

http://www.familysearch.org

Genealogy website managed by the LDS church to help members and nonmembers trace their family history. The church's copyrighted International Genealogical Index (IGI) database contains approximately 600 million individual names.

http://www.smgf.org

Sorenson Molecular Genealogy Foundation, formerly the BYU Molecular Genealogy Project headed by Professor Scott Woodward of the Department of Microbiology. The project aims to link family history with the genealogical information stored in DNA.

LDS Scholarship

http://www.sunstoneonline.com

The Sunstone Foundation encourages honest inquiry and respectful exchange among Mormons of all stripes, predominantly through its magazine, *Sunstone*. Since 1979 the foundation has held an annual conference in Salt Lake City.

http://www.dialoguejournal.com

Dialogue: A Journal of Mormon Thought is an independent quarterly "edited by Latter-day Saints who wish to bring their faith into dialogue with the larger stream of world religious thought" (May 2004).

http://byustudies.byu.edu

BYU Studies is an academic LDS journal that has been producing high-quality, peer-reviewed, LDS research since 1959.

Apologist Approaches to Mormonism

http://farms.byu.edu

The Foundation for Ancient Research and Mormon Studies (FARMS) employs professional defenders of the historicity of the Book of Mormon. FARMS is best known for its two major publications, the *Journal of Book of*

Mormon Studies and the *FARMS Review*. Most of the online resources are available only to members of FARMS, but there is a small free area.

http://www.fair-lds.org
The Foundation for Apologetic Information and Research (FAIR) is a non-profit organization dedicated to providing well-documented answers to criticisms of LDS doctrine, belief, and practice.

http://www.shields-research.org
The Scholarly and Historical Information Exchange for Latter-day Saints (SHIELDS) contains information to prevent people from being "deceived by those who would not tell you the truth about the LDS church" (May 2004). FAIR and SHIELDS are now affiliated.

http://www.jefflindsay.com
LDS member Jeff Lindsay is a self-proclaimed "Book of Mormon aficionado." His site is among the most frequently cited by Mormons, with "extensive and impressive evidences for the authenticity of the Book of Mormon as an ancient document dealing with real people and places, contrary to the endlessly and mindlessly repeated mantras of anti-Mormons." Lindsay has an extensive Book of Mormon-DNA page. In his view, the DNA studies that contradict the Book of Mormon are "wildly unscientific, though they are dressed in counterfeit robes of scientific objectivity" (May 2004).

http://www.ldscn.com/pearls
A site with an extensive set of links to other LDS websites.

Criticism of Mormonism

http://www.lds-mormon.com; http://www.2think.org
The LDS-Mormon and Honest Intellectual Inquiry websites are the brainchildren of Al Case, a former Mormon, who arguably has the largest and best collections of critical sources regarding Mormonism. In 2001 the sites received in excess of one million hits and over 700,000 full-page downloads per month.

http://home.teleport.com/~packham
Richard Packham left Mormonism in 1958 when he was twenty-five years old. His site contains an extensive array of material critical of Mormonism.

http://www.irr.org
The website of the Institute for Religious Research in Grand Rapids,

Michigan, contains hundreds of articles examining the history and doctrines of the LDS church.

http://www.exmormon.org

The Recovery from Mormonism site was founded in 1995 by former Mormon Eric Kettunen for those who are questioning the Mormon faith. In May 2004 the site contained over 250 stories written by ex-Mormons describing why they left the church. The site advocates no specific religious preference or religious activities after Mormonism. Story number 125 contains a personal account of my own departure from the Mormon church in 1998.

http://www.utlm.org

The famous Utah Lighthouse Ministry is operated by Jerald and Sandra Tanner, long-time critics of the Mormon church and known for their photographic reproductions and transcriptions of documents pertaining to LDS church history. Jerald is the great-great-grandson of John Tanner, who contributed considerable funds to Joseph Smith and the LDS church in 1835 when the church was deeply in debt. Sandra Tanner is a great-great-granddaughter of Brigham Young, the second president of the LDS church.

Ex-Mormon Support Groups

http://www.exmormonfoundation.org

"The Exmormon Foundation is a non-profit, non-sectarian organization dedicated to providing support to those who are leaving, or who have already left, the Mormon Church" (May 2004). The foundation sponsors the annual Exmormon Conference held in Salt Lake City each October.

http://www.exmormon.org/bboards.htm

An e-mail bulletin board located within the Recovery from Mormonism website described above. The bulletin board averaged 145,000 hits per day in 2003.

http://groups.yahoo.com/group/Exmormon

This ex-Mormon group "is maintained to provide a place where list members can discuss the problems and difficulties one may face by leaving the Mormon Church and adjusting to a life without Mormonism." In May 2004 there were about 1,000 list members generating between twenty and fifty posts per day.

http://groups.yahoo.com/group/XLDSWomen

An online support group for ex-Mormon women who discuss the aftershocks of the Mormon experience. In May 2004 there were about 500 list members generating between ten and forty posts per day.

http://www.irr.org/mit/mit-talk.html

A thoughtful site for ex-Mormon/Christian exchanges that bills itself as Mormons in Transition.

http://www.latterdaylampoon.com

A satirical website modeled on the official LDS site and operated by former Mormon Steven Clark.

Glossary of Genetics Terms

allele. One of the different forms of a particular gene occurring at the same position on homologous chromosomes.

amino acid. Organic molecules that serve as the building blocks of proteins.

autosome. Any chromosome other than the sex chromosomes.

base. Molecular units, also known as nucleotides, found in DNA: adenine (A), cytosine (C), guanine (G), and thymine (T).

bottleneck effect. Change in the gene pool of a surviving population after a dramatic reduction in the size of a parent population.

cell. The fundamental unit of living things. All organisms are made of cells.

chromosome. A thread-like structure found in the cell nucleus which contains a linear, end-to-end arrangement of genes.

codon. A section of DNA, three nucleotides in length, that codes for a particular amino acid.

crossing over. Reciprocal exchange of genetic material between homologous chromosomes.

DNA. Deoxyribonucleic acid—a double chain of linked bases which are the fundamental units of genes.

DNA lineage. A pedigree of related DNA-containing molecules, e.g., mitochondrial or Y chromosome DNAs.

DNA marker. Unique DNA sequences used to characterize or keep track of a gene, chromosome, or DNA lineage.

DNA sequence. The ordered arrangement of the bases within DNA.

eukaryote. An organism with cells containing a nucleus, e.g., fungi, plants, or animals.

fingerprint (DNA). A characteristic pattern of DNA fragments obtained from analysis of an individual's DNA.

founder effect. A change in the gene pool of a colonizing population because it is founded by a limited number of individuals from a parent population.

founding lineage (DNA). A DNA lineage present in the original founders of a population.

gene. The fundamental unit of heredity, a segment of DNA containing coded information for protein synthesis.

gene pool. The total collection of genes in a population.

genealogy. An account of the descent of a person or family through an ancestral line.

genetic code. The set of correspondences between base triplets (codons) in DNA and the amino acids in protein.

genetic distance. A measure of the relatedness between populations based on gene frequencies.

genetic drift. Changes in the gene pool of a small population due to chance.

genome. The entire complement of genetic material of an organism contained in its set of chromosomes.

haplogroup. A group of related haploid lineages (haplotypes), e.g., mitochondrial or Y chromosome DNA lineages.

haploid. Containing only one set of chromosomes.

haplotype. A haploid DNA lineage with a characteristic sequence.

homologous chromosomes. Chromosome pairs containing genes for the same traits at identical positions. One homologous chromosome is inherited from the father and the other from the mother.

lineage, *see* DNA lineage

marker, *see* DNA marker

maternal inheritance. Inherited solely from the mother, e.g., through the mitochondria.

messenger RNA, *see* mRNA

microsatellite. DNA markers comprising repeating units of DNA whose unit of repetition is usually two, three, or four bases.

mitochondria. Organelles in eukaryotic cells involved in energy metabolism.

mitochondrial DNA. Non-nuclear DNA contained within mitochondria.

molecular genetics. The study of the molecular processes underlying gene structure and function.

mRNA. Messenger ribonucleic acid is a single-stranded molecule similar to DNA that facilitates the flow of information in DNA for synthesizing proteins.

MtDNA, *see* mitochondrial DNA

mutation. A process that introduces changes into the order of bases in a DNA sequence.

nuclear DNA. DNA contained in chromosomes within the nucleus of the cell.

nucleotide, *see* base

nucleus. The chromosome-containing organelle in eukaryotic cells.

organelle. A body with a specialized function, found in eukaryotic cells.

paternal inheritance. Inherited solely from the father, e.g., through the Y chromosomes.

pedigree. A family tree drawn to show patterns of relatedness between characters or individuals.

polymorphism (DNA). DNA changes detected between two different DNA sequences.

polypeptide. A polymer (chain) of amino acids linked together.

prokaryote. An organism having cells lacking a nucleus, e.g., bacteria.

protein. An organic polymer constructed from chains of amino acids.

recombination. The formation of new genetic combinations in offspring by independent assortment and crossing over of parental chromosomes.

sex chromosome. A chromosome that plays a role in sex determination;

a chromosome whose presence or absence is correlated with the sex of the bearer.

transcription. The transfer of information from a DNA molecule into a messenger RNA molecule.

translation. The transfer of information from a messenger RNA molecule into a polypeptide involving a change of language from nucleic acids (RNA) to amino acids.

triplet. The three base pairs that comprise a codon.

X chromosome. The chromosome responsible for determining female sexual traits of an individual.

Y chromosome. The chromosome responsible for determining male sexual traits of an individual.

zygote. The diploid cell formed by the fusion of an egg and a sperm.

Works Cited

AAUP Committee on Academic Freedom and Tenure. 1998. "Report of Committee A," *Academe: Bulletin of the American Association of University Professors*. Sept./Oct: 71-4.

Adherents.com. 2003. *Largest Latter-day Saint Communities*, http://www.adherents.com/largecom/com_lds.html.

Adovasio, James M. 1983. "Evidence from Meadowcroft Rockshelter," in *Early Man in the New World*, ed. Richard J. Shutler. Beverly Hills: Sage Publications.

Alexeev, Valerii P., and Irena I. Gokhman. 1984. *Anthropology of the Asiatic Part of the USSR*. Moscow: IAPC (International Academic Publishing Co.) Nauka/Interperiodica.

Austin, Christopher C. 1999. "Lizards Took Express Train to Polynesia," *Nature* 397:113-14.

Axtell, James. 1981. *The European and the Indian: Essays in the Ethnohistory of Colonial North America*. Oxford: Oxford University Press.

Barker, Ian R. 1967. *The Connexion: The Mormon Church and the Maori People*, M.A. thesis. New Zealand: Victoria University of Wellington.

Bateman, Merrill J. 1997. "A Zion University and the Search for Truth," *Brigham Young Magazine*, address delivered at the BYU Annual University Conference. Winter.

Behar, Doron M., Michael F. Hammer, Daniel Garrigan, et al. 2004. "MtDNA Evidence for a Genetic Bottleneck in the Early History of the Ashkenazi Jewish Population," *European Journal of Human Genetics* 12:355-64.

Bellwood, Peter S. 1979. *Man's Conquest of the Pacific: The Prehistory of Southeast Asia and Oceania.* New York: Oxford University.

__________. 1987. *The Polynesians: Prehistory of an Island People,* rev. ed. London: Thames and Hudson.

Berkhofer, Robert F. 1978. *The White Man's Indian: Images of the American Indian from Columbus to the Present.* New York: Knopf.

Bolnick, Daniel A. 2003. "Genetic Relationships among the Prehistoric Adena and Hopewell," Seventy-second Meeting of the American Association of Physical Anthropologists, visual display.

Bonatto, Sandro Luis, Alan J. Redd, Francisco Mauro Salzano, et al. 1996. "Lack of Ancient Polynesian-Amerindian Contact," *American Journal of Human Genetics* 59:253-58.

Bonatto, Sandro Luis, and Francisco M. Salzano. 1997. "Diversity and Age of the Four Major MtDNA Haplogroups and Their Implications for the Peopling of the New World," *American Journal of Human Genetics* 61: 1413-23.

Bonné-Tamir, Batsheva, J. G. Bodmer, W. F. Bodmer, et al. 1978. "HLA Polymorphisms in Israel: An Overall Comparative Analysis," *Tissue Antigens* 11:235-50.

Bonné-Tamir, Batsheva, M. J. Johnson, A. Natali, et al. 1986. "Human Mitochondrial DNA Types in Two Israeli Populations: A Comparative Study at the DNA Level," *American Journal of Human Genetics* 38:341-51.

Bortolini, Maria-Catira, Francisco M. Salzano, Mark G. Thomas, et al. 2003. "Y-Chromosome Evidence for Differing Ancient Demographic Histories in the Americas," *American Journal of Human Genetics* 73:524-39.

Brewerton, Ted E. 1995. "The Book of Mormon: A Sacred Ancient Record," *Ensign* 25:30-31.

Bright, John. 1981. *A History of Israel*, 3rd ed. Philadelphia: Westminster.

Brodie, Fawn M. 1971. *No Man Knows My History: The Life of Joseph Smith, the Mormon Prophet,* 2nd ed. New York: Knopf.

Brooke, John Hedley. 1991. "Science and Religion: Some Historical Perspectives," in *Cambridge History of Science.* Cambridge: Cambridge University Press.

Brown, Michael D., Sayed H. Hosseini, Antonio Torroni, et al. 1998. "MtDNA Haplogroup X: An Ancient Link between Europe/Western Asia and North America?" *American Journal of Human Genetics* 63:1852-61.

Cann, Rebecca L., Mark Stoneking, and Allan C. Wilson. 1987. "Mitochondrial DNA and Human Evolution," *Nature* 325:31-36.

Cavalli-Sforza, L. Luca, Paolo Menozzi, and Joanna Mountain. 1988. "Reconstruction of Human Evolution: Bringing Together Genetic, Archaeological, and Linguistic Data," *Proceedings of the National Academy of Sciences of the United States of America* 85:6002-06.

Cavalli-Sforza L. Luca, Paolo Menozzi, and Alberto Piazza. 1994. *The History and Geography of Human Genes*. Princeton, New Jersey: Princeton University Press.

Chadwick, Bruce A., and T. Garrow. 1992. "Native Americans," in the *Encyclopedia of Mormonism,* ed. Daniel H. Ludlow. New York: Macmillan.

Clark, John E. 1992. "Book of Mormon Geography," in the *Encyclopedia of Mormonism,* ed. Daniel H. Ludlow. New York: Macmillan.

Clement, Russell T. 1980. "Polynesian Origins: More Word on the Mormon Perspective," *Dialogue: A Journal of Mormon Thought* 13:88-98.

Coe, Michael D. 1973. "Mormons and Archaeology: An Outside View," *Dialogue: A Journal of Mormon Thought* 8:40-48.

Coe, Michael D., Dean Snow, and Elizabeth Benson. 1986. *Atlas of Ancient America.* Oxford: Equinox.

Coe, Sophie D., and Michael D. Coe. 1996. *The True History of Chocolate.* London: Thames and Hudson.

Collingwood, Robin G. 1946. *The Idea of History*. Oxford: Clarendon Press.

Crawford, Michael H. 1998. *The Origins of Native Americans: Evidence from Anthropological Genetics*. Cambridge: Cambridge University Press.

Darwin, Charles. 1859. *On the Origin of Species by Means of Natural Selection: Or the Preservation of Favoured Races in the struggle for Life*. London: John Murray.

Dennell, Robin. 1983. *European Economic Prehistory: A New Approach*. London: Academic Press.

Derenko, Miroslava V., Tomasz Grzybowski, Boris A. Malyarchuk, et al. 2001. "The Presence of Mitochondrial Haplogroup X in Altaians from South Siberia," *American Journal of Human Genetics* 69:237-41.

Diamond, Jared M. 1997. *Guns, Germs, and Steel: The Fates of Human Societies*. New York: W. W. Norton.

__________. 1988. "Express Train to Polynesia," *Nature* 336:307-08.

Dillehay, Tom D. 1997. *Monte Verde, A Late Pleistocene Settlement in Chile:*

Vol. 2, The Archaeological Context and Interpretation. Washington D. C.: Smithsonian Institution Press.

DNA vs. the Book of Mormon video documentary. 2002. Brigham City, Utah: Living Hope Ministries.

Doebley, John F., Major M. Goodman, and Charles W. Stuber. 1984. "Isoenzymatic Variation in Zea (Gramineae)," *Systematic Botany* 9:203-18.

Douglas, Norman. 1974. "The Sons of Lehi and the Seed of Cain: Racial Myths in Mormon Scripture and Their Relevance to the Pacific Islands, *Journal of Religious History* 8:90-104.

Egan, Dan. 2000a. "At BYU: Banking on Blood for Genetic History," *Salt Lake Tribune*, October 15.

__________. 2000b. "BYU Gene Data May Shed Light on Origin of Book of Mormon's Lamanites," *Salt Lake Tribune*, November 30.

Encyclopedia Britannica. 2004. Sv. "Canaan," *Encyclopedia Britannica Online*.

Encyclopedia Judaica. 1972. Jerusalem: Keter Publishing.

Eyring, Henry. 1998. *Reflections of a Scientist*. Salt Lake City: Deseret Book.

Fagan, Brian M. 1987. *The Great Journey: The Peopling of Ancient America*. New York: Thames and Hudson.

Ferguson, Thomas Stuart. 1958. *One Fold and One Shepherd*. San Francisco: Books of California.

Flannery, Tim F. 1994. *The Future Eaters: An Ecological History of the Australasian Lands and People*. Chatswood, New South Wales: Reed Books.

Foley, Robert. 1998. "The Context of Human Genetic Evolution," *Genome Research* 8:339-47.

Forster, Peter, Rosalind Harding, Antonio Torroni, et al. 1996. "Origin and Evolution of Native American MtDNA Variation: A Reappraisal," *American Journal of Human Genetics* 59:935-45.

Gibbons, Francis M. 1995. *Spencer W. Kimball: Resolute Disciple, Prophet of God*. Salt Lake City: Deseret Book.

Giddings, James L. 1962. "Development of Tree-Ring Dating as an Archaeological Aid," in *Tree Growth*, ed. Theodore T. Kozlowski. New York: Ronald Press Co.

Gonzáles, Ana M., Antonio Brehm, José A. Pérez, et al. 2003. "Mitochondrial DNA Affinities at the Atlantic Fringe of Europe," *American Journal of Physical Anthropology* 120:391-404.

Gordon, Tamar. 1988. *Inventing Mormon Identity in Tonga,* Ph.D. diss., University of California at Berkeley.

Gospel Principles. 1997. Salt Lake City: Church of Jesus Christ of Latter-day Saints.

Gould, Stephen Jay. 1999. *Rock of Ages: Science and Religion in the Fullness of Life.* New York: Ballantine.

Gray, Russell D., and Fiona M. Jordan. 2000. "Language Trees Support the Express-Train Sequence of Austronesian Expansion," *Nature* 405: 1052-55.

Grayson, Donald K. 1987. "Death by Natural Causes," *Natural History* 5:8-13.

Green, Dee F. 1973. "Book of Mormon Archaeology: The Myths and the Alternatives," *Dialogue: A Journal of Mormon Thought* 4:71-80.

Greenberg, Joseph H., Christy G. Turner, and Stephen L. Zegura. 1986. "Settlement of the Americas: A Comparison of the Linguistic, Dental, and Genetic Evidence," *Current Anthropology* 27:477-97.

Griffin, James B. 1979. "The Origin and Dispersion of American Indians in North America," in *The First Americans: Origins, Affinities, and Adaptations,* eds. William S. Laughlin and Albert B. Harper. New York: Gustav Fischer.

Guidon, Niède, and Georgette Delibrias. 1986. "Carbon-14 Dates Point to Man in the Americas 32,000 Years Ago," *Nature* 321:769-71.

Hammer, Michael F., Tatiana Karafet, Arani Rasanayagam, et al. 1998. "Out of Africa and Back Again: Nested Cladistic Analysis of Human Y Chromosome Variation," *Molecular Biology and Evolution* 15:427-41.

Hammer, Michael F., Alan J. Redd, Elizabeth T. Wood, et al. 2000. "Jewish and Middle Eastern Non-Jewish Populations Share a Common Pool of Y-chromosome Biallelic Haplotypes," *Proceedings of the National Academy of Sciences of the USA* 97:6769-74.

Hammer, Michael F., and Stephen L. Zegura. 2002. "The Human Y Chromosome Haplogroup Tree: Nomenclature and Phylogeography of Its Major Divisions," *Annual Review of Anthropology* 31:303-21.

Hauswirth, William W., C. D. Dickel, D. J. Rowold, et al. 1994. "Inter- and Intra-population Studies of Ancient Humans," *Experientia* 50:585-91.

Hertzberg, Mark, K. M. Mickleson, S. W. Serjeantson, et al. 1989. "An Asian-Specific 9-bp Deletion of Mitochondrial DNA Is Frequently Found in Polynesians," *American Journal of Human Genetics* 44:504-10.

Heun, Manfred, Ralf Schäfer-Pregl, Dieter Klawan, et al. 1997. "Site of Einkorn Wheat Domestication Identified by DNA Fingerprinting," *Science* 278:1312-14.

Heyerdahl, Thor. 1950. *Kon-Tiki: Across the Pacific by Raft,* transl. F. H. Lyon. Chicago: Rand McNally.

Hinckley, Gordon B. 1998. "Gordon Hinckley: Distinguished Religious Leader of the Mormons," interviewed by Larry King, CNN (Cable News Network), September 8.

Holland, Jeffrey R. 2001. "Witnesses unto Me," *Ensign* 31:14-16.

Hopkins, David M. 1979. "Landscape and Climate of Beringia during Late Pleistocene and Holocene Time," in *The First Americans: Origins, Affinities, and Adaptations,* eds. William S. Laughlin and Albert B. Harper. New York: Gustav Fischer.

Hunter, Milton R. 1956. *Archaeology and the Book of Mormon.* Salt Lake City: Deseret Book.

__________. 1970. *Great Civilizations and the Book of Mormon,* 3 vols. Salt Lake City: Bookcraft.

Hunter, William A. 1978. "History of the Ohio Valley," in *Handbook of North American Indians,* ed. William C. Sturtevant, vol. 15. Washington, D.C.: Smithsonian Institution.

Huoponen, Kirsi, Antonio Torroni, Patricia R. Wickman, et al. 1997. "Mitochondrial DNA and Y Chromosome-Specific Polymorphisms in the Seminole Tribe of Florida," *European Journal of Human Genetics* 5:25-34.

Hurles, Matthew E., Catherine Irven, Jayne Nicholson, et al. 1998. "European Y-Chromosomal Lineages in Polynesians: A Contrast to the Population Structure Revealed by MtDNA," *American Journal of Human Genetics* 63:1793-806.

Hurles, Matthew E., Emma Maund, Jayne Nicholson, et al. 2003. "Native American Y Chromosomes in Polynesia: The Genetic Impact of the Polynesian Slave Trade," *American Journal of Human Genetics* 72:1282-7.

Hurles, Matthew E., Jayne Nicholson, Elena Bosch, et al. 2002. "Y Chromosomal Evidence for the Origins of Oceanic-Speaking Peoples," *Genetics* 160:289-303.

Ingman, Max, Henrik Kaessmann, Svante Pääbo, et al. 2000. "Mitochondrial Genome Variation and the Origin of Modern Humans," *Nature* 408: 708-13.

Ingstad, Anne S. 1985. *The Norse Discovery of America,* trans. Elizabeth S. Seeberg. Oslo: Norwegian University Press.

Jackson, Kent P. 1988. Review of Hugh W. Nibley, *Old Testament and Related Studies,* vol. 1 of *The Collected Works of Hugh Nibley*, ed. Gary Gillum. *BYU Studies* 28:114.

Jackson, Lionel E., and Alejandra Duk-Rodkin. 1996. "Quarternary Geology of the Ice-Free Corridor: Glacial Controls on the Peopling of the New World," in *Prehistoric Mongoloid Dispersal,* eds. Takeru Akazawa and Emoke J. Szathmáry. Oxford: Oxford University Press.

Jeffery, Duane E. 1973. "Seers, Savants, and Evolution: The Uncomfortable Interface," *Dialogue: A Journal of Mormon Thought* 8:41-75.

Jennings, Francis. 1993. *The Founders of America: How Indians Discovered the Land, Pioneered in It, and Created Great Classical Civilizations.* New York: W. W. Norton.

Jessee, Dean C., ed. 1984. *The Personal Writings of Joseph Smith.* Salt Lake City: Deseret Book.

Johnson, Cooper. 2003. "DNA and the Book of Mormon," *The Foundation for Apologetic Information & Research.* http://www.fairlds.org/apol/bom/bom01.html.

Johnson, David J. 1992. "Archaeology," in *Encyclopedia of Mormonism*, ed. Daniel H. Ludlow. New York: Macmillan.

Jones, Steve. 1996. *In the Blood: God, Genes, and Destiny*. London: HarperCollins.

Kaessmann, Henrik, Florian Heissig, Arndt von Haeseler, et al. 1999. "DNA Sequence Variation in a Non-Coding Region of Low Recombination on the Human X Chromosome," *Nature Genetics* 22:78-81.

Karafet, Tatiana M., Stephen L. Zegura, Olga Posukh, et al. 1999. "Ancestral Asian Source(s) of New World Y-Chromosome Founder Haplotypes," *American Journal of Human Genetics* 64:817-31.

Karafet, Tatiana, Stephen L. Zegura, Jennifer Vuturo-Brady, et al. 1997. "Y Chromosome Markers and Trans-Bering Strait Dispersals," *American Journal of Physical Anthropology* 102:301-14.

Kayser, Manfred, Silke Brauer, Gunter Weiss, et al. 2000. "Melanesian Origin of Polynesian Y Chromosomes," *Current Biology* 10:1237-46.

Kenney, Scott G. 1997. "Mormons and the Smallpox Epidemic of 1853," *The Hawaiian Journal of History* 31:1-26.

Kimball, Spencer W. 1960. "The Day of the Lamanites," *The Improvement Era,* December, 922-23.

__________. 1975. "First Presidency Message: Our Paths Have Met Again," *Ensign* 5:2-7.

__________. 1982. *The Teachings of Spencer W. Kimball, Twelfth President of the Church of Jesus Christ of Latter-day Saints.* Salt Lake City: Bookcraft.

King, Jonathan C. 1999. *First Peoples, First Contacts: Native Peoples of North America.* Cambridge, Mass.: Harvard University Press.

Krings, Matthias, Anne Stone, Ralf W. Schmitz, et al. 1997. "Neanderthal DNA Sequences and the Origin of Modern Humans," *Cell* 90:19-30.

Lalueza, Carles, A. Pérez-Pérez, E. Prats, et al. 1997. "Lack of Founding Amerindian Mitochondrial DNA Lineages in Extinct Aborigines from Tierra del Fuego-Patagonia," *Human Molecular Genetics* 6:41-6.

Larson, Charles M. 1992. *By His Own Hand upon Papyrus: A New Look at the Joseph Smith Papyri.* Rev. ed. Grand Rapids, Mich.: Institute for Religious Research.

Larson, Stan. 1996. *Quest for the Gold Plates: Thomas Stuart Ferguson's Archaeological Search for the Book of Mormon.* Salt Lake City: Freethinker Press.

Lell, Jeffrey T., Michael D. Brown, Theodore G. Schurr, et al. 1997. "Y Chromosome Polymorphisms in Native American and Siberian Populations: Identification of Native American Y Chromosome Haplotypes," *Human Genetics* 100:536-43.

Leonard, Jennifer A., Robert K. Wayne, Jane Wheeler, et al. 2002. "Ancient DNA Evidence for Old World Origin of New World Dogs," *Science* 298:1613-16.

Libby, Willard F. 1955. *Radiocarbon Dating,* 2nd ed. Chicago: University of Chicago Press.

Lindberg, David C., and Ronald L. Numbers. 1986. *In God and Nature: Historical Essays on the Encounter between Christianity and Science*, eds. David C. Lindberg and Ronald L. Numbers. Berkeley: University of California Press.

Lindsay, Jeff D. "Does DNA Evidence Refute the Book of Mormon?" *LDS FAQ: Frequently Asked Questions about Latter-day Saint Beliefs,* http://www.jefflindsay.com/LDSFAQ/DNA.shtml.

Lorenz, Joseph G., and David G. Smith. 1996. "Distribution of Four Founding MtDNA Haplogroups among Native North Americans," *American Journal of Physical Anthropology* 101:307-23.

Ludlow, Daniel H., ed. 1992. *Encyclopedia of Mormonism: The History, Scripture, Doctrine, and Procedure of the Church of Jesus Christ of Latter-day Saints,* 5 vols. New York: Macmillan. Volume 5 contains the Book of Mormon, Doctrine and Covenants, and Pearl of Great Price—the three additional scriptures of the Mormon church.

Lum, J. Koji, and Rebecca Luisa Cann. 1998. "MtDNA and Language Support a Common Origin of Micronesians and Polynesians in Island Southeast Asia," *American Journal of Physical Anthropology* 105:109-19.

Lum, J. Koji, and Rebecca Luisa Cann. 2000. "MtDNA Lineage Analyses: Origins and Migrations of Micronesians and Polynesians," *American Journal of Physical Anthropology* 113:151-68.

Lum, J. Koji, Rebecca Luisa Cann, Jeremy J. Martinson, et al. 1998. "Mitochondrial and Nuclear Genetic Relationships among Pacific Island and Asian Populations," *American Journal of Human Genetics* 63:613-24.

Lum, J. Koji, Olga Rickards, C. Ching, et al. 1994. "Polynesian Mitochondrial DNAs Reveal Three Deep Maternal Lineage Clusters," *Human Biology* 66:567-90.

Lyon, Eugene, and Bob Sacha. 1992. "Search for Columbus," *National Geographic* 181:2-39.

Macaulay, Vincent, Martin Richards, Eileen Vega, et al. 1999. "The Emerging Tree of West Eurasian MtDNAs: A Synthesis of Control-Region Sequences and RFLPs," *American Journal of Human Genetics* 64:232-49.

Madsen, Truman G., and John W. Welch. 1985. "Did B. H. Roberts Lose Faith in the Book of Mormon?" Provo, Utah: FARMS Preliminary Report.

Matheny, Ray T. 1984. "Book of Mormon Archaeology," an address delivered at the Sunstone Theological Symposium. Provo, Utah: Special Collections, Harold B. Lee Library, Brigham Young University.

Matisoo-Smith, Elizabeth, R. M. Roberts, G. J. Irwin, et al. 1998. "Patterns of Prehistoric Human Mobility in Polynesia Indicated by MtDNA from the Pacific Rat," *Proceedings of the National Academy of Sciences of the USA* 95:15145-50.

McClellan, David A. 2003. "Detecting Lehi's Genetic Signature: Possible, Probable, or Not?" *FARMS Review* 15:35-90.

McConkie, Bruce R. 1979. *Mormon Doctrine*, 2nd ed. Salt Lake City: Bookcraft.

McEvedy, Colin, and Richard Jones. 1978. *Atlas of World Population History*. New York: Penguin Books.

McGlone, Matthew S., Atholl J. Anderson, and Richard N. Holdaway. 1994. "An Ecological Approach to the Polynesian Settlement of New Zealand," in *The Origins of the First New Zealanders*, ed. Douglas G. Sutton. Auckland: Auckland University Press.

McGraw, Douglas J. 2000. "Andrew Ellicott Douglass and the Big Trees," *American Scientist* 88:440-47.

McNickle, D'Arcy. 1971. "Americans Called Indians," in *North American Indians in Historical Perspective,* ed. Eleanor B. Leacock and Nancy O. Lurie. New York: Random House.

Meldrum, D. Jeffrey, and Trent D. Stephens. 2003. "Who Are the Children of Lehi?" *Journal of Book of Mormon Studies* 12:38-51.

Melton, Terry, R. Peterson, A. J. Redd, et al. 1995. "Polynesian Genetic Affinities with Southeast Asian Populations as Identified by MtDNA Analysis," *American Journal of Human Genetics* 57:403-14.

Metcalfe, Brent L. 1993. *New Approaches to the Book of Mormon: Explorations in Critical Methodology*. Salt Lake City: Signature Books.

__________. 2004. "Reinventing Lamanite Identity," *Sunstone* 131:20-25.

Mills, Lisa Ann. 2003. "Mitochondrial DNA Analysis of the Ohio Hopewell of the Hopewell Mound Group," Ph.D. thesis, Ohio State University.

"Mistakes in the News: DNA and the Book of Mormon." 2003. *Official Internet Site of the Church of Jesus Christ of Latter-day Saints,* http://www.lds.org/newsroom/mistakes.

Murphy, Thomas W. 2002. "Lamanite Genesis, Genealogy, and Genetics," in *American Apocrypha: Essays on the Book of Mormon*, ed. by Dan Vogel and Brent Metcalfe. Salt Lake City: Signature Books.

Murphy, Thomas W., and Simon G. Southerton. 2003. "Genetic Research: A 'Galileo Event' for Mormons," *Anthropology News* 44:20.

Murray-McIntosh, Rosalind P., Brian J. Scrimshaw, Peter J. Hatfield, et al. 1998. "Testing Migration Patterns and Estimating Founding Population Size in Polynesia by Using Human MtDNA Sequences," *Proceedings of the National Academy of Sciences of the United States of America* 95:9047-52.

Ndoro, Webber. 1997. "Great Zimbabwe," *Scientific American*. November: 62-67.

Neel, James V., Robert J. Biggar, and Rem I. Sukernik. 1994. "Virologic and Genetic Studies Relate Amerind Origins to the Indigenous People of the Mongolia/Manchuria/Southeastern Siberia Region," *Proceedings of the National Academy of Sciences of the United States of America* 91:10737-41.

Nelkin, Dorothy. 1982. *The Creation Controversy: Science or Scripture in the Schools*. New York: W. W. Norton.

Neves, Walter, and Hector Pucciarelli. 1998. "The Zhoukoudian Upper Cave Skull 101 as Seen from the Americans," *Journal of Human Evolution* 34:219-22.

Nibley, Hugh. 1946. *No, Ma'am, That's Not History: A Brief Review of Mrs. Brodie's Reluctant Vindication of a Prophet She Seeks to Expose*. Salt Lake City: Bookcraft.

__________. 1952. "Letter to Professor 'F'," *Improvement Era*, qtd. in vol. 5, *The Collected Works of Hugh Nibley: Lehi in the Desert, The World of the Jaredites, There Were Jaredites*, eds. John W. Welch, Darrell L. Matthews, and Stephen R. Callister. Salt Lake City: Deseret Book and the Foundation for Ancient Research and Mormon Studies, 1988.

__________. 1964. *An Approach to the Book of Mormon*, 2nd ed. Salt Lake City: Deseret Book.

__________. 1988. *Lehi in the Desert: The World of the Jaredites; There Were Jaredites*, eds. John W. Welch, Darrell L. Matthews, and Stephen R. Callister. Salt Lake City: Deseret Book and the Foundation for Ancient Research and Mormon Studies.

Oaks, Dallin H. 1993. "The Historicity of the Book of Mormon." Featured Papers, *Foundation for Ancient Research and Mormon Studies*, http://farms.byu.edu/publications/papers.php.

Okladnikov, Alexei P. 1964. "Ancient Population of Siberia and Its Culture," in *The Peoples of Siberia*, eds. Maksim G. Levin and Leonid P. Potapov. Chicago: University of Chicago Press.

Olsen, Kenneth M., and Barbara A. Schaal. 1999. "Evidence on the Origin of Cassava: Phylogeography of *Manihot esculenta*," *Proceedings of the National Academy of Sciences of the United States of America* 96:5586-91.

Ostling, Richard N., and Joan K. Ostling. 1999. *Mormon America: The Power and the Promise*. San Francisco: HarperSanFrancisco.

Packer, Boyd K. 1981. "The Mantle is Far, Far Greater than the Intellect," *BYU Studies* 21:259.

__________. 1992. "Our Moral Environment," *Ensign*, Proceedings of the 162nd Semi-annual General Conference of The Church of Jesus Christ of Latter-day Saints. May: 66.

Parfitt, Tudor. 1997. *Journey to the Vanished City: The Search for a Lost Tribe of Israel*, rev. and enlgd. London: Phoenix.

Parry, Donald W. 1998. "The Flood and the Tower of Babel," *Ensign* 28:35-41.

Pena, Sérgio D., Fabrício R. Santos, Néstor O. Bianchi, et al. 1995. "A Major Founder Y-Chromosome Haplotype in Amerindians," *Nature Genetics* 11:15-16.

Persuitte, David. 2000. *Joseph Smith and the Origins of The Book of Mormon*, 2nd ed. Jefferson, North Carolina: McFarland & Company.

Petersen, Mark E. 1954. Race Problems as They Affect the Church: Address at the Convention of Teachers of Religion on the College Level, August 27, Brigham Young University. Ms. 376, Special Collections, J. Willard Marriott Library, University of Utah, Salt Lake City.

__________. 1962. *Conference Report: The One Hundred Thirty-Second Annual General Conference of the Church of Jesus Christ of Latter-day Saints*, April, 112.

Peterson, Daniel C. 2000. "Mounting Evidence for the Book of Mormon," *Ensign* 30:18-24.

__________. 2003a. "Of Galileo Events, Hype, and Suppression: Or Abusing Science and Its History," *FARMS Review* 15:ix-lx

Popov, A. A., and B. O. Dolgikh. 1964. "The Kets," in *The Peoples of Siberia*, eds. Maksim G. Levin and Leonid P. Potapov. Chicago: University of Chicago Press.

Powell, Joseph F., and Walter A. Neves. 1999. "Craniofacial Morphology of the First Americans: Pattern and Process in the Peopling of the New World," *Yearbook of Physical Anthropology* 42:153-88.

Pratt, Addison. 1990. *The Journals of Addison Pratt: Being a Narrative of Yankee Whaling in the Eighteen Twenties, a Mormon Mission to the Society Islands, and of Early California and Utah in the Eighteen Forties and Fifties*, ed. S. George Ellsworth. Salt Lake City: University of Utah Press.

Preston, Douglas. 1997. "The Lost Man," *The New Yorker*, 16 June, 70-81.

Price, A. Grenfell, ed. 1958. *Explorations of Captain James Cook in the Pacific as Told by Selections of His Own Journals, 1768-1779.* New York: Heritage Press.

Prokof'yeva, E. D. 1964. "The Selkups," in *The Peoples of Siberia,* eds. Maksim G. Levin and Leonid P. Potapov. Chicago: University of Chicago Press.

Ramenofsky, Ann F. 1987. *Vectors of Death: The Archaeology of European Contact.* Albuquerque: University of New Mexico Press.

Reidla, Maere, Toomas Kivisild, Ene Metspalu, et al. 2003. "Origin and Diffusion of MtDNA Haplogroup X," *American Journal of Human Genetics* 73:1178-90.

Reynolds, Noel B. 1999. "The Coming Forth of the Book of Mormon in the Twentieth Century," *BYU Studies* 38:6-47.

Ribeiro dos Santos, Andrea Kely, Sidney Manuel Santos, Ana Lúcia Machado, et al. 1996. "Heterogeneity of Mitochondrial DNA Haplotypes in Pre-Columbian Natives of the Amazon Region," *American Journal of Physical Anthropology* 101:29-37.

Richards, Martin, Melna Côrte-Real, Peter Forster, et al. 1996. "Paleolithic and Neolithic Lineages in the European Mitochondrial Gene Pool," *American Journal of Human Genetics* 59:185-203.

Richards, Martin, Vincent Macaulay, Eileen Hickey, et al. 2000. "Tracing European Founder Lineages in the Near Eastern MtDNA Pool," *American Journal of Human Genetics* 67:1251-76.

Roberts, Brigham H. 1992. *Studies of the Book of Mormon*, 2nd ed. Salt Lake City: Signature Books.

Roberts, Richard G., R. Jones, and M. A. Smith. 1990. "Thermoluminescence Dating of a 50,000-Year-Old Human Occupation Site in Northern Australia," *Nature* 345:153-56.

Roosevelt, Ana C., Marcondes Lima da Costa, Cristiane Lopes Machado, et al. 1996. "Paleoindian Cave Dwellers in the Amazon: The Peopling of the Americas," *Science* 272:373-84.

Roper, Matthew. 2003a. "Nephi's Neighbors: Book of Mormon Peoples and Pre-Columbian Populations," *FARMS Review* 15: 91-128.

__________. 2003b. "Swimming in the Gene Pool: Israelite Kinship Relations, Genes, and Genealogy," *FARMS Review* 15: 129-64.

Rosser, Zoë H., Tatiana Zerjal, Matthew Edward Adojaan, et al. 2000.

"Y-Chromosomal Diversity in Europe Is Clinal and Influenced Primarily by Geography Rather than by Language," *American Journal of Human Genetics* 67:1526-43.

Ryan, William B. F., Walter C. Pitman III, Candace O. Major, et al. 1997. "An Abrupt Drowning of the Black Sea Shelf," *Marine Geology* 138: 119-26.

Santos, Fabrício R., Arpita Pandya, Chris Tyler-Smith, et al. 1999. "The Central Siberian Origin for Native American Y Chromosomes," *American Journal of Human Genetics* 64:619-28.

Santos, Sidney E., Andrea K. Ribeiro dos Santos, Daniel Meyer, et al. 1996. "Multiple Founder Haplotypes of Mitochondrial DNA in Amerindians Revealed by RFLP and Sequencing," *Annals of Human Genetics* 60: 305-19.

Sauer, Jonathan D. 1993. *Historical Geography of Crop Plants: A Selected Roster*. Boca Raton: CRC Press.

Savolainen, Peter, Ya-ping Zhang, Jing Luo, et al. 2002. "Genetic Evidence for an East Asian Origin of Domestic Dogs," *Science* 298:1610-13.

Schurr, Theodore G. 2000. "Mitochondrial DNA and the Peopling of the New World," *American Scientist* 46:246-53.

Schurr, Theodore G., Scott W. Ballinger, Yik Y. Gan, et al. 1990. "Amerindian Mitochondrial DNAs Have Rare Asian Mutations at High Frequencies Suggesting They Derived from Four Primary Maternal Lineages," *American Journal of Human Genetics* 46:613-23.

Scozzari, Rosaria, Fulvio Cruciani, Piero Santolamazza, et al. 1997. "MtDNA and Y Chromosome-Specific Polymorphisms in Modern Ojibwa: Implications about the Origin of Their Gene Pool," *American Journal of Human Genetics* 60:241-44.

Seielstad, Mark, Nadira Yuldasheva, Nadia Singh, et al. 2003. "A Novel Y-Chromosome Variant Puts an Upper Limit on the Timing of First Entry into the Americas," *American Journal of Human Genetics* 73:700-05.

Seife, Charles. 2001. "Papal Science: Science and Religion Advance Together at Pontifical Academy," *Science* 291:1472-74.

Silva, Wilson A., Sandro L. Bonatto, Adriano J. Holanda, et al. 2002. "Mitochondrial Genome Diversity of Native Americans Supports a Single Early Entry of Founder Populations into America," *American Journal of Human Genetics* 71:187-92.

Silverberg, Robert. 1968. *Mound Builders of Ancient America: The Archaeology of a Myth*. Athens, Ohio: Ohio University Press.

Smith, Bruce D. 1994. *The Emergence of Agriculture*. New York: Scientific American Library.

Smith, David G., Ripan S. Malhi, Jason Eshleman, et al. 1999. "Distribution of MtDNA Haplogroup X among Native North Americans," *American Journal of Physical Anthropology* 110:271-84.

Smith, George D. 1984. "Is There Any Way to Escape These Difficulties?: The Book of Mormon Studies of B. H. Roberts," *Dialogue: A Journal of Mormon Thought* 17:94-111.

Smith, Joseph. 1932. *History of the Church of Jesus Christ of Latter-day Saints*, ed. B. H. Roberts, 7 vols. Salt Lake City: The Church of Jesus Christ of Latter-day Saints.

Solheim, Wilhelm G. 1970. "Northern Thailand, Southeast Asia, and World History," *Asian Perspectives* 13:145-57.

Soodyall, Himla, Trefor Jenkins, and Mark Stoneking. 1995. "'Polynesian' MtDNA in the Malagasy," *Nature Genetics* 10:377-78.

Sorenson, John L. 1966. "Some Voices from the Dust." Review of *Papers of the Fifteenth Annual Symposium on the Archaeology of the Scriptures*, in *Dialogue: A Journal of Mormon Thought* 1:144-49.

__________. 1985. *An Ancient American Setting for the Book of Mormon*. Salt Lake City: Deseret Book.

__________. 1992. "When Lehi's Party Arrived in the Land, Did They Find Others There?" *Journal of Book of Mormon Studies* 1:1-34.

__________. 1995. "The Book of Mormon in Ancient America," FARMS Book of Mormon Lecture Series typescript. Provo, Utah: Foundation for Ancient Research and Mormon Studies.

__________. 2000a. "New Light: Genetics Indicates that Polynesians Were Connected to Ancient America," *Journal of Book of Mormon Studies* 9:1.

__________. 2000b. "New Light: The Problematic Role of DNA Testing in Unraveling Human History," *Journal of Book of Mormon Studies* 9:2.

__________. 2002. "An Interview with John L. Sorenson," *Journal of Book of Mormon Studies* 11:80-85.

Sorenson, John L., and Matthew Roper. 2003. "Before DNA," *Journal of Book of Mormon Studies* 12:6-23.

Soustelle, Jacques. 1984. *The Olmecs: The Oldest Civilization in Mexico*. New York: Doubleday.

Southerton, Simon G. 2000. "DNA Genealogies of American Indians and the Book of Mormon," *Recovery from Mormonism* website, http://www.exmormon.org/whylft125.htm.

__________. 2002. "DNA Genealogies of Native Americans and Polynesians," address given at Exmormon Foundation Annual Conference, October.

Spriggs, Matthew. 1985. "The Lapita Cultural Complex," in *Out of Asia: Peopling the Americas and the Pacific*, eds. Robert L. Kirk and Emöke Szathmary. Canberra: Australian National University Press.

__________. 1996. "What is Southeast Asian about Lapita?" in *Prehistoric Mongoloid Dispersals*, eds. Takeru Akazawa and Emöke Szathmary. Oxford: Oxford University Press.

Starikovskaya, Yelena B., Rem I. Sukernik, Theodore G. Schurr, et al. 1998. "MtDNA Diversity in Chukchi and Siberian Eskimos: Implications for the Genetic History of Ancient Beringia and the Peopling of the New World," *American Journal of Human Genetics* 63:1473-91.

Stone, Anne C., and Mark Stoneking. 1998. "MtDNA Analysis of a Prehistoric Oneota Population: Implications for the Peopling of the New World," *American Journal of Human Genetics* 62:1153-70.

Stoneking, Mark, and Allan C. Wilson. 1989. "Mitochondrial DNA," in *The Colonization of the Pacific: A Genetic Trail*, eds. Adrian V. S. Hill and Susan W. Serjeantson. Oxford: Oxford University Press.

Strahler, Arthur N. 1987. *Science and Earth History: The Evolution/Creation Controversy*. Buffalo, New York: Prometheus Books.

Straus, Lawrence G. 1989. "Age of the Modern Europeans," *Nature* 342: 476-77.

Stephens, John Lloyd. 1841. *Incidents of Travel in Central America: Chiapas and Yucatan*. New York: Harper.

Su, Bing, Li Jin, Peter Underhill, et al. 2000. "Polynesian Origins: Insights from the Y Chromosome," *Proceedings of the National Academy of Sciences of the United States of America* 97:8225-28.

Sykes, Bryan, and Catherine Irven. 2000. "Surnames and the Y Chromosome," *American Journal of Human Genetics* 66:1417-19.

Sykes, Bryan, A. Leiboff, J. Low-Beer, et al. 1995. "The Origins of the Polynesians: An Interpretation from Mitochondrial Lineage Analysis," *American Journal of Human Genetics* 57:1463-75.

Szeinberg, Amir. 1979. "Polymorphic Evidence for a Mediterranean Origin of the Ashkenazi Community," in *Genetic Diseases among Ashkenazi Jews*, eds. Richard Merle Goodman and Arno G. Motulsky. New York: Raven.

Talmage, James E. 1915. *Jesus the Christ: A Study of the Messiah and His Mission according to Holy Scriptures both Ancient and Modern*. Salt Lake City: The Church of Jesus Christ of Latter-day Saints.

Taylor, Peter. 1990. *The Atlas of Australian History*. Frenchs Forest, New South Wales: Child and Associates.

Terrell, John E. 1986. *Prehistory in the Pacific Islands*. Cambridge: Cambridge University Press.

Thomas, Mark D. 1999. *Digging in Cumorah: Reclaiming Book of Mormon Narratives*. Salt Lake City: Signature Books.

Thomas, Mark G., Tudor Parfitt, Deborah Weiss, et al. 2000. "Y Chromosomes Traveling South: The Cohen Modal Haplotype and the Origins of the Lemba 'Black Jews of Southern Africa,'" *American Journal of Human Genetics* 66:674-86.

Thomas, Mark G., Karl Skorecki, Haim Ben-Ami, et al. 1998. "Origins of Old Testament Priests," *Nature* 394:138-40.

Thomas, Mark G., Michael E. Weale, Abigail L. Jones, et al. 2002. "Founding Mothers of Jewish Communities: Geographically Separated Jewish Groups Were Independently Founded by Very Few Female Ancestors," *American Journal of Human Genetics* 70:14111-20.

Thorne, Alan. 1980. "The Arrival of Man in Australia," in *The Cambridge Encyclopedia of Archaeology*, ed. Andrew Sherratt. New York: Crown Publishers.

Thorne, Alan, Rainer Grün, Graham Mortimer, et al. 1999. "Australia's Oldest Human Remains: Age of the Lake Mungo 3 Skeleton," *Journal of Human Evolution* 36:591-612.

Thornton, Russell. 1987. *American Indian Holocaust and Survival: A Population History Since 1492*. Norman, Oklahoma: University of Oklahoma Press.

Tikochinski, Yaron, U. Ritte, S. R. Gross, et al. 1991. "MtDNA Polymorphism in Two Communities of Jews," *American Journal of Human Genetics* 48:129-36.

Torroni, Antonio, Kirsi Huoponen, Paolo Francalacci, et al. 1996. "Classification of European MtDNAs from an Analysis of Three European Populations," *Genetics* 144:1835-50.

Torroni, Antonio, Y. S. Chen, O. Semino, et al. 1994a. "MtDNA and Y-Chromosome Polymorphisms in Four Native American Populations from Southern Mexico," *American Journal of Human Genetics* 54:303-18.

Torroni, Antonio, James V. Neel, Ramiro Barrantes, et al. 1994b. "Mitochondrial DNA 'Clock' for the Amerinds and Its Implications for Timing Their Entry into North America," *Proceedings of the National Academy of Sciences of the United States of America* 91:1158-62.

Torroni, Antonio, J. A. Miller, L. G. Moore, et al. 1994c. "Mitochondrial DNA Analysis in Tibet: Implications for the Origin of the Tibetan Population and Its Adaptation to High Altitude," *American Journal of Physical Anthropology* 93:189-99.

Torroni, Antonio, M. T. Lott, M. F. Cabell, et al. 1994d. "MtDNA and the Origin of Caucasians: Identification of Ancient Caucasian-Specific Haplogroups, One of Which is Prone to a Recurrent Somatic Duplication in the D-Loop Region," *American Journal of Human Genetics* 55:760-76.

Torroni, Antonio, Theodore G. Schurr, Charles C. Yang, et al. 1992. "Native American Mitochondrial DNA Analysis Indicates that the Amerind and the Nadene Populations Were Founded by Two Independent Migrations," *Genetics* 130:153-62.

Trigger, Bruce G. 1978. "Early Iroquoian Contacts with Europeans," in *Handbook of North American Indians,* ed. William C. Sturtevant, vol. 15. Washington, D.C.: Smithsonian Institution.

Turner, Christy G. 1983. "Dental Evidence for the Peopling of the Americas," in *Early Man in the New World,* ed. Richard J. Shutler. Beverly Hills: Sage Publications.

Underhill, Peter A., Lin Jin, Rachel Zemans, et al. 1996. "A Pre-Columbian Y Chromosome-Specific Transition and Its Implications for Human Evolutionary History," *Proceedings of the National Academy of Sciences of the United States of America* 93:196-200.

Underwood, Grant. 2000. "Mormonism, the Maori and Cultural Authenticity," *Journal of Pacific History* 35:133-46.

Valladas, Hélène, H. Cachier, and P. Maurice. 1992. "Direct Radiocarbon Dates for Prehistoric Paintings at the Altamira, El Castillo, and Niaux Caves," *Nature* 357:68-70.

Vogel, Dan. 1986. *Indian Origins and the Book of Mormon: Religious Solutions from Columbus to Joseph Smith.* Salt Lake City: Signature Books.

Vogel, Dan. 1994. "The Locations of Joseph Smith's Early Treasure Quests," *Dialogue: A Journal of Mormon Thought* 27:197-231.

Wahlgren, Erik. 1986. *The Vikings and America.* London and New York: Thames and Hudson.

Wang, Rong-Lin, Adrian Stec, Jody Hey, et al. 1999. "The Limits of Selection during Maize Domestication," *Nature* 398:236-39.

Warren, Bruce W. 1990. Review of F. Richard Hauck, *Deciphering the Geography of the Book of Mormon: Settlements and Routes in Ancient America,* and John L. Sorenson, *An Ancient American Setting for the Book of Mormon,* in *BYU Studies* 30:127.

Whiting, Michael F. 2003a. "DNA and the Book of Mormon: A Phylogenetic Perspective," *Journal of Book of Mormon Studies* 12:24-35.

__________. 2003b. "Does DNA Evidence Refute the Authenticity of the Book of Mormon? Responding to the Critics," lecture presented at Brigham Young University.

Whittington, Stephen L., and David M. Reed, eds. 1997. *Bones of the Maya: Studies of Ancient Skeletons.* Washington, D. C.: Smithsonian Institution.

Willey, Gordon R., and Jeremy A. Sabloff. 1980. *A History of American Archaeology,* 2nd ed. San Francisco: W. H. Freeman and Company.

Williams, Robert C., A. G. Steinberg, et al. 1985. "GM Allotypes in Native Americans: Evidence for Three Distinct Migrations across the Bering Land Bridge," *American Journal of Physical Anthropology* 66:1-19.

Wilson, James. 1999. *The Earth Shall Weep: A History of Native America.* New York: Atlantic Monthly Press.

Woodward, Scott R. 2001. "DNA and the Book of Mormon," presented at the 3rd Annual Mormon Apologetics Conference, Foundation for Apologetic Information & Research, http://www.fairlds.org.

Wurm, Stephen A. 1967. "Linguistics and the Prehistory of the Southwestern Pacific," *Journal of Pacific History* 2:25-38.

YCC (Y Chromosome Consortium). 2002. "A Nomenclature System for the Tree of Human Y-Chromosomal Binary Haplogroups," *Genome Research* 12:339-48.

Zegura, Stephen L., Tatiana M. Karafet, Lev A. Zhivotovsky, et al. 2004.

"High Resolution SNPs and Microsatellite Haplotypes Point to a Single, Recent Entry of Native American Y Chromosomes into the Americas," *Molecular Biology and Evolution* 21:164-75.

Zhang, Dapeng, Jim Cervantes, Zosimo Huaman, et al. 2000. "Assessing Genetic Diversity of Sweet Potato—*Ipomoea batatas* (L.) Lam.—Cultivars from Tropical America Using AFLP," *Genetic Resources and Crop Evolution* 47:659-65.

Zhang, Dapeng, Marc Ghislain, Zosimo Huaman, et al. 1998. "RAPD Variation in Sweet Potato—*Ipomoea batatas* (L.) Lam.—Cultivars from South America and Papua New Guinea," *Genetic Resources and Crop Evolution* 45:271-77.

Zhang, Dapeng P., Genoveva Rossel, Albert Kriegner, et al. 2004. "From Latin America to Oceania: The Historic Dispersal of Sweet Potato Re-examined Using AFLP," *Genetic Resources and Crop Evolution* 51:115-20.

Index

A

B

C